Albert Einstein
Hendrik Antoon Lorentz

ON THE THEORY
OF RELATIVITY

Edited by Vesselin Petkov

MINKOWSKI
Institute Press

Albert Einstein (1879-1955)
Hendrik Antoon Lorentz (1853-1928)

Published 2021

ISBN: 978-1-989970-73-7 (softcover)
ISBN: 978-1-989970-74-4 (ebook)

Minkowski Institute Press
Montreal, Quebec, Canada
http://minkowskiinstitute.org/mip/

For information on all Minkowski Institute Press
publications visit our website at
http://minkowskiinstitute.org/mip/books/

EDITOR'S PREFACE

This collection contains new publications of three articles by Einstein

- "What is the Theory Of Relativity"

- "A Brief Outline of the Development of the Theory of Relativity"

- "The Problem of Space, Ether and the Field in Physics"

and two small books:

- A. Einstein, *Sidelights on Relativity**

- H. A. Lorentz, *The Einstein Theory of Relativity: A Concise Statement.*†

A number of Editor's notes (as footnotes) are added in the text mostly to provide additional clarification in order to avoid confusion or misconceptions.

The texts were typeset in LaTeX by Svetla Petkova and noticed typos were corrected.

---

*A. Einstein, *Sidelights on Relativity*. Translated by G. B. Jeffery and W. Perrett (Methuen & Co. ltd., London 1922).

†H. A. Lorentz, *The Einstein Theory of Relativity: A Concise Statement*, 3rd ed. (Brentano's Publishers, New York 1920). Originally published in the *Nieuwe Rotterdamsche Courant* of November 19, 1919.

i

27 December 2021

Vesselin Petkov
Minkowski Institute
Montreal

CONTENTS

WHAT IS THE THEORY OF RELATIVITY

Published in *The London Times*, November 28, 1919, translated by R. W. Lawson

I gladly accede to the request of your colleague to write something for The Times (London) on relativity. After the lamentable breakdown of the old active intercourse between men of learning, I welcome this opportunity of expressing my feelings of joy and gratitude toward the astronomers and physicists of England. It is thoroughly in keeping with the great and proud traditions of scientific work in your country that eminent scientists should have spent much time and trouble, and your scientific institutions have spared no expense, to test the implications of a theory which was perfected and published during the war in the land of your enemies. Even though the investigation of the influence of the gravitational field of the sun on light rays is a purely objective matter, I cannot forbear to express my personal thanks to my English colleagues for their work; for without it I could hardly have lived to see the most important implication of my theory tested.

We can distinguish various kinds of theories in physics. Most of them are constructive. They attempt to build up a picture of the more complex phenomena out of the materials of a relatively simple formal scheme from which they start out. Thus the kinetic theory of gases seeks to reduce mechanical, thermal, and diffusional processes to movements of molecules – i.e., to build them up out of the hypothesis of molecular motion. When we say that we have succeeded in understanding a group of natural processes, we invariably mean that a constructive theory has been found which covers the processes in question.

Along with this most important class of theories there exists a second, which I will call "principle-theories." These employ the analytic, not the synthetic, method. The elements which form their basis and starting-point are not hypothetically constructed but empirically discovered ones, general characteristics of natural processes, principles that give rise to mathematically formulated criteria which the separate processes or the theoretical representations of them have to satisfy. Thus the science of thermodynamics seeks by analytical means to deduce necessary conditions, which separate events have to satisfy, from the universally experienced fact that perpetual motion is impossible.

The advantages of the constructive theory are completeness, adaptability, and clearness, those of the principle theory are logical perfection and security of the foundations.

The theory of relativity belongs to the latter class. In order to grasp its nature, one needs first of all to become acquainted with the principles on which it is based. Before I go into these, however, I must observe that the theory of relativity resembles a building consisting of two separate stories, the special theory and the general theory. The special theory, on which the general theory rests, applies to all physical phenomena with the exception of gravitation; the general theory provides the law of gravitation and its relations to the other forces of nature.

It has of course been known since the days of the ancient Greeks that in order to describe the movement of a body, a second body is needed to which the movement of the first is referred. The movement of a vehicle is considered in reference to the earth's surface; that of a planet to the totality of the visible fixed stars. In physics the body to which events are spatially referred is called the coordinate system. The laws of the mechanics of Galileo and Newton, for instance, can only be formulated with the aid of a coordinate system.

The state of motion of the coordinate system may not, however, be arbitrarily chosen, if the laws of mechanics are to be valid (it must be free from rotation and acceleration). A coordinate system which is admitted in mechanics is called an "inertial system." The state of motion of an inertial system is according to mechanics not one that is determined uniquely by nature. On the contrary, the following definition holds good: a coordinate system that is moved uniformly and in a straight line relative to an inertial system is likewise an inertial system. By the "special principle of relativity" is meant the generalization of this definition to include any natural event whatever: thus, every universal law of nature which is valid in relation to a coordinate system C, must also be valid, as it stands, in relation to a coordinate system C', which is in uniform translatory motion relatively to C.

The second principle, on which the special theory of relativity rests, is the "principle of the constant velocity of light *in vacuo*." This principle asserts that light *in vacuo* always has a definite velocity of propagation (independent of the state of motion of the observer or of the source of the light). The confidence which physicists place in this principle springs from the successes achieved by the electrodynamics of Maxwell and Lorentz.

Both the above-mentioned principles are powerfully supported by experience, but appear not to be logically reconcilable. The special theory of relativity finally succeeded in reconciling them logically by a modification of kinematics – i.e., of the doctrine of the laws relating to space and time (from the point of view of physics). It became clear that to speak of the simultaneity of two events had no meaning except in relation to a given coordinate system, and that the shape of measuring devices and the speed at which clocks move depend on their state of motion with respect to the coordinate system.

But the old physics, including the laws of motion of Galileo and Newton, did not fit in with the suggested relativist kinematics. From the latter, general mathematical conditions issued, to which natural laws had to conform, if the above-mentioned two principles were really to apply. To these, physics had to be adapted. In particular, scientists arrived at a new law of motion for (rapidly moving) mass points, which was admirably confirmed in the case of electrically charged particles. The most important upshot of the special theory of relativity concerned the inert masses of corporeal systems. It turned out that the inertia of a system necessarily depends on its energy-content, and this led straight to the notion that inert mass is simply latent energy. The principle of the conservation of mass lost its independence and became fused with that of the conservation of energy.

The special theory of relativity, which was simply a systematic development of the electrodynamics of Maxwell and Lorentz, pointed beyond itself, however.

Should the independence of physical laws of the state of motion of the coordinate system be restricted to the uniform translatory motion of coordinate systems in respect to each other? What has nature to do with our coordinate systems and their state of motion? If it is necessary for the purpose of describing nature, to make use of a coordinate system arbitrarily introduced by us, then the choice of its state of motion ought to be subject to no restriction; the laws ought to be entirely independent of this choice (general principle of relativity).

The establishment of this general principle of relativity is made easier by a fact of experience that has long been known, namely, that the weight and the inertia of a body are controlled by the same constant (equality of inertial and gravitational mass). Imagine a coordinate system which is rotating uniformly with respect to an inertial system in the Newtonian manner. The centrifugal forces which manifest themselves in relation to this system must, according to Newton's teaching, be regarded as effects of inertia. But these centrifugal forces are, exactly like the forces of gravity, proportional to the masses of the bodies. Ought it not to be possible in this case to regard the coordinate system as stationary and the centrifugal forces as gravitational forces? This seems the obvious view, but classical mechanics forbid it.

This hasty consideration suggests that a general theory of relativity must supply the laws of gravitation and the consistent following up of the idea has justified our hopes.

But the path was thornier than one might suppose, because it demanded the abandonment of Euclidean geometry. This is to say, the laws according to which solid bodies may be arranged in space do not completely accord with the spatial laws attributed to bodies by Euclidean geometry. This is what we mean when we talk of the "curvature of space." The fundamental concepts of the "straight line," the "plane," etc., thereby lose their precise significance in physics.

In the general theory of relativity the doctrine of space and time, or kinematics, no longer figures as a fundamental independent of the rest of physics. The geometrical behavior of bodies and the motion of clocks rather depend on gravitational fields, which in their turn are produced by matter.

The new theory of gravitation diverges considerably, as regards principles, from Newton's theory. But its practical results agree so nearly with those of Newton's theory that it is difficult to find criteria for distinguishing them which are accessible to experience. Such have been discovered so far:

1. In the revolution of the ellipses of the planetary orbits round the sun (confirmed in the case of Mercury).

2. In the curving of light rays by the action of gravitational fields (confirmed by the English photographs of eclipses).

3. In a displacement of the spectral lines toward the red end of the spectrum in the case of light transmitted to us from stars of considerable magnitude (unconfirmed so far).[1]

---

[1]EDITOR'S NOTE: This effect had been repeatedly verified later. There had been some initial observations of the gravitational redshift in spectral lines from starts before 1960. But the first definite confirmation of this prediction of general relativity was done by a terrestrial experiment using the Mössbauer effect by Pound and Rebka in 1960 (R. V. Pound and G. A. Rebka, Jr., "Apparent Weight of Photons" *Phys. Rev. Lett.* **4** (1960), 337-341).

The chief attraction of the theory lies in its logical completeness. If a single one of the conclusions drawn from it proves wrong, it must be given up; to modify it without destroying the whole structure seems to be impossible.

Let no one suppose, however, that the mighty work of Newton can really be superseded by this or any other theory. His great and lucid ideas will retain their unique significance for all time as the foundation of our whole modern conceptual structure in the sphere of natural philosophy.

Note: Some of the statements in your paper concerning my life and person owe their origin to the lively imagination of the writer. Here is yet another application of the principle of relativity for the delectation of the reader: today I am described in Germany as a "German savant," and in England as a "Swiss Jew." Should it ever be my fate to be represented as a *bête noire,* I should, on the contrary, become a "Swiss Jew" for the Germans and a "German savant" for the English.

A Brief Outline of the Development of the Theory of Relativity

Published in *Nature*, **106** (No. 2677); February 17, 1921; pp. 782-784, translated by R. W. Lawson

There is something attractive in presenting the evolution of a sequence of ideas in as brief a form as possible, and yet with a completeness sufficient to preserve throughout the continuity of development. We shall endeavour to do this for the Theory of Relativity, and to show that the whole ascent is composed of small, almost self-evident steps of thought.

The entire development starts off from, and is dominated by, the idea of Faraday and Maxwell, according to which all physical processes involve a continuity of action (as opposed to action at a distance), or, in the language of mathematics, they are expressed by partial differential equations. Maxwell succeeded in doing this for electro-magnetic processes in bodies at rest by means of the conception of the magnetic effect of the vacuum-displacement-current, together with the postulate of the identity of the nature of electrodynamic fields produced by induction, and the electrostatic field.

The extension of electrodynamics to the case of moving bodies fell to the lot of Maxwell's successors.

H. Hertz attempted to solve the problem by ascribing to empty space (the aether) quite similar physical properties to those possessed by ponderable matter; in particular, like ponderable matter, the aether ought to have at every point a definite velocity. As in bodies at rest, electromagnetic or magneto-electric induction ought to be determined by the rate of change of the electric or magnetic flow respectively, provided that these velocities of alteration are referred to surface elements moving with the body. But the theory of Hertz was opposed to the fundamental experiment of Fizeau on the propagation of light in flowing liquids. The most obvious extension of Maxwell's theory to the case of moving bodies was incompatible with the results of experiment.

At this point, H. A. Lorentz came to the rescue. In view of his unqualified adherence to the atomic theory of matter, Lorentz felt unable to regard the latter as the seat of continuous electromagnetic fields. He thus conceived of these fields as being conditions of the aether, which was regarded as continuous. Lorentz considered the aether to be intrinsically independent of matter, both from a mechanical and a physical point of view. The aether did not take part in the motions of matter, and a reciprocity between aether and matter could be assumed only in so far as the latter was considered to be the carrier of attached electrical charges. The great value of the theory of Lorentz lay in the fact that the entire electrodynamics of bodies at rest and of bodies in motion was led back to Maxwell's equations of empty space. Not only did this theory surpass that of Hertz from the point of view of method, but with its aid H. A. Lorentz was also pre-eminently successful in explaining the experimental facts.

The theory appeared to be unsatisfactory only in

one point of fundamental importance. It appeared to give preference to one system of coordinates of a particular state of motion (at rest relative to the aether) as against all other systems of coordinates in motion with respect to this one. In this point the theory seemed to stand in direct opposition to classical mechanics, in which all inertial systems which are in uniform motion with respect to each other are equally justifiable as systems of coordinates (Special Principle of Relativity). In this connection, all experience also in the realm of electrodynamics (in particular Michelson's experiment) supported the idea of the equivalence of all inertial systems, i.e. was in favour of the special principle of relativity.

The Special Theory of Relativity owes its origin to this difficulty, which, because of its fundamental nature, was felt to be intolerable. This theory originated as the answer to the question: Is the special principle of relativity really contradictory to the field equations of Maxwell for empty space? The answer to this question appeared to be in the affirmative. For if those equations are valid with reference to a system of coordinates K, and we introduce a new system of coordinates K' in conformity with the — to all appearances readily establishable — equations of transformation

$$\left. \begin{array}{l} x' = x - vt \\ y' = y \\ z' = z \\ t' = t \end{array} \right\} \text{(Galileo transformation)},$$

then Maxwell's field equations are no longer valid in the new coordinates (x', y', z', t'). But appearances are deceptive. A more searching analysis of the physical significance of space and time rendered it evident that the Galileo transformation is founded on arbitrary as-

11

sumptions, and in particular on the assumption that the statement of simultaneity has a meaning which is independent of the state of motion of the system of co-ordinates used. It was shown that the field equations for *vacuo* satisfy the special principle of relativity, provided we make use of the equations of transformation stated below:

$$\left.\begin{array}{l} x' = \dfrac{x-vt}{\sqrt{1-v^2/c^2}} \\[2ex] y' = y \\ z' = z \\[1ex] t' = \dfrac{t-vx/c^2}{\sqrt{1-v^2/c^2}} \end{array}\right\} \quad \text{(Lorentz transformation)}.$$

In these equations x, y, z represent the coordinates measured with measuring-rods which are at rest with reference to the system of coordinates, and t represents the time measured with suitably adjusted clocks of identical construction, which are in a state of rest.

Now in order that the special principle of relativity may hold, it is necessary that all the equations of physics do not alter their form in the transition from one inertial system to another, when we make use of the Lorentz transformation for the calculation of this change. In the language of mathematics, all systems of equations that express physical laws must be covariant with respect to the Lorentz transformation. Thus, from the point of view of method, the special principle of relativity is comparable to Carnot's principle of the impossibility of perpetual motion of the second kind, for, like the latter, it supplies us with a general condition which all natural laws must satisfy.

Later, H. Minkowski found a particularly elegant and suggestive expression for this condition of covariance, one which reveals a formal relationship between

Euclidean geometry of three dimensions and the space-time continuum of physics.

Euclidean Geometry of Three Dimensions	*Special Theory of Relativity*
Corresponding to two neighbouring points in space, there exists a numerical measure (distance ds) which conforms to the equation	Corresponding to two neighbouring points in space-time (point events), there exists a numerical measure (distance ds) which conforms to the equation
$$ds^2 = dx_1^2 + dx_2^2 + dx_3^2.$$	$$ds^2 = dx_1^2 + dx_2^2 + dx_3^2 + dx_4^2.$$
It is independent of the system of coordinates chosen, and can be measured with the unit measuring-rod.	It is independent of the inertial system chosen, and can be measured with the unit measuring-rod and a standard clock. x_1, x_2, x_3, are here rectangular coordinates, whilst $x_4 = \sqrt{-1}ct$ is the time multiplied by the imaginary unit and by the velocity of light.
The permissible transformations are of such a character that the expression for ds is invariant, i.e. the linear orthogonal transformations are permissible.	The permissible transformations are of such a character that the expression for ds is invariant, i.e. those linear orthogonal substitutions are permissible which maintain the semblance of reality of x_1, x_2, x_3, x_4. These substitutions are the Lorentz transformations.
With respect to these transformations, the laws of Euclidean geometry are invariant.	With respect to these transformations, the laws of physics are invariant.

From this it follows that, in respect of its *role* in

13

the equations of physics, though not with regard to its physical significance, time is equivalent to the space coordinates (apart from the relations of reality). From this point of view, physics is, as it were, a Euclidean geometry of four dimensions, or, more correctly, a statics in a four-dimensional Euclidean continuum.

The development of the special theory of relativity consists of two main steps, namely, the adaptation of the space-time "metrics" to Maxwell's electrodynamics, and an adaptation of the rest of physics to that altered space-time "metrics." The first of these processes yields the relativity of simultaneity, the influence of motion on measuring-rods and clocks, a modification of kinematics, and in particular a new theorem of addition of velocities. The second process supplies us with a modification of Newton's law of motion for large velocities, together with information of fundamental importance on the nature of inertial mass.

It was found that inertia is not a fundamental property of matter, nor, indeed, an irreducible magnitude, but a property of energy. If an amount of energy E be given to a body, the inertial mass of the body increases by an amount E/c^2, where c is the velocity of light *in vacuo*. On the other hand, a body of mass m is to be regarded as a store of energy of magnitude mc^2.

Furthermore, it was soon found impossible to link up the science of gravitation with the special theory of relativity in a natural manner. In this connection I was struck by the fact that the force of gravitation possesses a fundamental property, which distinguishes it from electro-magnetic forces. All bodies fall in a gravitational field with the same acceleration, or — what is only another formulation of the same fact — the gravitational and inertial masses of a body are numerically equal to each other. This numerical equality suggests identity in character. Can gravitation and inertia be identical? This question leads directly to the General Theory of Relativity. Is it not possible for me to regard the earth as free from rotation, if I conceive of the centrifugal force, which acts on all bodies at rest relatively to the earth, as being a "real" field of gravitation, or part of such a field? If this idea can be carried out, then we shall have proved in very truth the identity of gravitation and inertia. For the same property which is regarded as *inertia* from the point of view of a system not taking part in the rotation can be interpreted as *gravitation* when considered with respect to a system that shares the rotation. According to Newton, this interpretation is impossible, because by Newton's law the centrifugal field cannot be regarded as being produced by matter, and because in Newton's theory there is no place for a "real" field of the "Koriolis-field" type. But perhaps Newton's law of field could be replaced by another that fits in with the field which holds with respect to a "rotating" system of coordinates? My conviction of the identity of inertial and gravitational mass aroused within me the feeling of absolute confidence in the correctness of this interpretation. In this connection I gained encouragement from the following idea. We are familiar with the "apparent" fields which are valid relatively to systems of coordinates possessing arbitrary motion with respect to an inertial system. With the aid of these special fields we should be able to study the law which is satisfied in general by gravitational fields. In this connection we shall have to take account of the fact that the ponderable masses will be the determining factor in producing the field, or, according to the fundamental result of the special theory of relativity, the energy density — a magnitude having the transformational character of a tensor.

On the other hand, considerations based on the metrical results of the special theory of relativity led to the result that Euclidean metrics can no longer be valid with respect to accelerated systems of coordinates. Although it retarded the progress of the theory several years, this enormous difficulty was mitigated by our knowledge that Euclidean metrics holds for small domains. As a consequence, the magnitude ds, which was physically defined in the special theory of relativity hitherto, retained its significance also in the general theory of relativity. But the coordinates themselves lost their direct significance, and degenerated simply into numbers with no physical meaning, the sole purpose of which was the numbering of the space-time points. Thus in the general theory of relativity the coordinates perform the same function as the Gaussian coordinates in the theory of surfaces. A necessary consequence of the preceding is that in such general coordinates the measurable magnitude ds must be capable of representation in the form

$$ds^2 = \sum_{uv} g_{uv} dx_u dx_v,$$

where the symbols g_{uv} are functions of the space-time coordinates. From the above it also follows that the nature of the space-time variation of the factors g_{uv} determines, on one hand the space time metrics, and on the other the gravitational field which governs the mechanical behaviour of material points.

The law of the gravitational field is determined mainly by the following conditions: First, it shall be valid for an arbitrary choice of the system of coordinates; secondly, it shall be determined by the energy tensor of matter; and thirdly, it shall contain no higher differential coefficients of the factors g_{uv} than the second, and must be linear in these. In this way a law was obtained which, although fundamentally different from Newton's law, corresponded so exactly to the latter in the deductions derivable from it that only very few criteria were to be found on which the theory could be decisively tested by experiment.

The following are some of the important questions which are awaiting solution at the present time. Are electrical and gravitational fields really so different in character that there is no formal unit to which they can be reduced? Do gravitational fields play a part in the constitution of matter, and is the continuum within the atomic nucleus to be regarded as appreciably non-Euclidean? A final question has reference to the cosmological problem. Is inertia to be traced to mutual action with distant masses? And connected with the latter: Is the spatial extent of the universe finite? It is here that my opinion differs from that of Eddington. With Mach, I feel that an affirmative answer is imperative, but for the time being nothing can be proved. Not until a dynamical investigation of the large systems of fixed stars has been performed from the point of view of the limits of validity of the Newtonian law of gravitation for immense regions of space will it perhaps be possible to obtain eventually an exact basis for the solution of this fascinating question.

The Problem of Space, Ether and the Field in Physics

Published in Albert Einstein, *Essays in science*, translated by Alan Harris (Philosophical Library, New York 1934) pp. 61-77

Scientific thought is a development of pre-scientific thought. As the concept of space was already fundamental in the latter, we must begin with the concept of space in pre-scientific thought. There are two ways of regarding concepts, both of which are necessary to understanding. The first is that of logical analysis. It answers the question. How do concepts and judgments depend on each other? In answering it we are on comparatively safe ground. It is the security by which we are so much impressed in mathematics. But this security is purchased at the price of emptiness of content. Concepts can only acquire content when they are connected, however indirectly, with sensible experience. But no logical investigation can reveal this connection; it can only be experienced. And yet it is this connection that determines the cognitive value of systems of concepts.

Take an example. Suppose an archaeologist belonging to a later culture finds a text-book of Euclidean geometry without diagrams. He will discover how the words "point," "straight-line," "plane" are used in the propositions. He will also see how the latter are deduced from each other. He will even be able to frame new propositions according to the known rules. But the framing of these propositions will remain an empty word-game for him, as long as "point," "straight-line," "plane," etc., convey nothing to him. Only when they do convey something will geometry possess any real content for him. The same will be true of analytical mechanics, and indeed of any exposition of the logically deductive sciences.

What does this talk of "straight-line," "point," "intersection," etc., conveying something to one, mean? It means that one can point to the parts of sensible experience to which those words refer. This extra-logical problem is the essential problem, which the archaeologist will only be able to solve intuitively, by examining his experience and seeing if he can discover anything which corresponds to those primary terms of the theory and the axioms laid down for them. Only in this sense can the question of the nature of a conceptually presented entity be reasonably raised.

With our pre-scientific concepts we are very much in the position of our archaeologist in regard to the ontological problem. We have, so to speak, forgotten what features in the world of experience caused us to frame those concepts, and we have great difficulty in representing the world of experience to ourselves without the spectacles of the old-established conceptual interpretation. There is the further difficulty that our language is compelled to work with words which are inseparably connected with those primitive concepts. These are the obstacles which confront us when we try to describe the essential nature of the pre-scientific concept of space.

One remark about concepts in general, before we turn to the problem of space: concepts have reference to sensible experience, but they are never, in a logical sense, deducible from them. For this reason I have never been able to understand the quest of the *a priori* in the Kantian sense. In any ontological question, the only possible procedure is to seek out these characteristics in the complex of sense experiences to which the concepts refer.

Now as regards the concept of space: this seems to presuppose the concept of the solid object. The nature of the complexes and sense-impressions which are probably responsible for that concept has often been described. The correspondence between certain visual and tactile impressions, the fact that they can be continuously followed out through time, and that the impressions can be repeated at any movement (taste, sight), are some of those characteristics. Once the concept of the solid object is formed in connection with the experiences just mentioned – which concept by no means presupposes that of space or spatial relation – the desire to get an intellectual grasp of the relations of such solid bodies is bound to give rise to concepts which correspond to their spatial relations. Two solid objects may touch one another or be distant from one another. In the latter case, a third body can be inserted between them without altering them in any way, in the former not. These spatial relations are obviously real in the same sense as the bodies themselves. If two bodies are of equal value for the filling of one such interval, they will also prove of equal value for the filling of other intervals. The interval is thus shown to be independent of the selection of any special body to fill it; the same is universally true of spatial relations. It is plain that this independence, which is a principle condition of the usefulness of framing purely geometrical concepts, is not necessary *a priori.* In my opinion, this concept of the interval, detached as it is from the selection of any special body to occupy it, is the starting point of the whole concept of space.

Considered, then, from, the point of view of sense experience, the development of the concept of space seems, after these brief indications, to conform to the following schema – solid body; spatial relations of solid bodies; interval; space. Looked at in this way, space appears as something real in the same sense as solid bodies.

It is clear that the concept of space as a real thing already existed in the extra-scientific conceptual world. Euclid's mathematics, however, knew nothing of this concept as such; they confined themselves to the concepts of the object, and the spatial relations between objects. The point, the plane, the straight line, length, are solid objects idealized. All spatial relations are reduced to those of contact (the intersection of straight lines and planes, points lying on straight lines, etc.). Space as a continuum does not figure in the conceptual system at all. This concept was first introduced by Descartes, when he described the point-in-space by its coordinates. Here for the first time geometrical figures appear, up to a point, as parts of infinite space, which is conceived as a three-dimensional continuum.

The great superiority of the Cartesian treatment of space is by no means confined to the fact that it applies analysis to the purposes of geometry. The main point seems rather to be this: The geometry of the Greeks prefers certain figures (the straight line, the plane) in geometrical descriptions; other figures (e.g., the ellipse) are only accessible to it because it constructs or defines them with the help of the point, the straight line and the plane. In the Cartesian treatment on the other hand, all surfaces are, in principle, equally represented, without any arbitrary preference for linear figures in the construction of geometry.

In so far as geometry is conceived as the science of laws governing the mutual relations of practically rigid bodies in space, it is to be regarded as the oldest branch of physics. This science was able, as I have already observed, to get along without the concept of space as such, the ideal corporeal forms – point, straight line, plane, length – being sufficient for its needs. On the other hand, space as a whole, as conceived by Descartes, was absolutely necessary to Newtonian physics. For dynamics cannot manage with the concepts of the mass point and the (temporally variable) distance between mass points alone. In Newton's equations of motion the concept of acceleration plays a fundamental part, which cannot be defined by the temporally variable intervals between points alone. Newton's acceleration is only thinkable or definable in relation to space as a whole. Thus to the geometrical reality of the concept of space a new inertia-determining function of space was added. When Newton described space as absolute, he no doubt meant this real significance of space, which made it necessary for him to attribute to it a quite definite state of motion, which yet did not appear to be fully determined by the phenomena of mechanics. This space was conceived as absolute in another sense also; its inertia-determining effect was conceived as autonomous, i.e., not to be influenced by any physical circumstance whatever; it affected masses, but nothing affected it.

And yet in the minds of physicists space remained until the most recent time simply the passive container of all events, playing no part in physical happenings itself. Thought only began to take a new turn with the wave theory of light and the theory of the electromagnetic field of Faraday and Clerk Maxwell. It became clear that there existed in free space conditions which propagated themselves in waves, as well as localized fields which were able to exert force on electrical masses or magnetic poles brought to the spot. Since it would have seemed utterly absurd to the physicists of the nineteenth century to attribute physical functions or states to space itself, they invented a medium pervading the whole of space, on the model of ponderable matter – the ether, which was supposed to act as a vehicle for electromagnetic phenomena, and hence for those of light also. The states of this medium, imagined as constituting the electromagnetic fields, were at first thought of mechanically, on the model of the elastic deformations of rigid bodies. But this mechanical theory of the ether was never quite successful and so the idea of a closer explanation of the nature of the etheric fields was given up. The ether thus became a kind of matter whose only function was to act as a substratum for electrical fields which were by their very nature not further analyzable. The picture was, then, as follows: Space is filled by the ether, in which the material corpuscles or atoms of ponderable matter swim; the atomic structure of the latter had been securely established by the turn of the century.

Since the reciprocal action of bodies was supposed to be accomplished through fields, there had also to be a gravitational field in the ether, whose field-law had, however, assumed no clear form at that time. The ether was only accepted as the seat of all operations of force which make themselves effective across space. Since it had been realized that electrical masses in motion produce a magnetic field, whose energy acted as a model for inertia, inertia also appeared as a field-action localized in the ether.

The mechanical properties of the ether were at first a mystery. Then came H. A. Lorentz's great discovery. All the phenomena of electromagnetism then known could be explained on the basis of two assumptions: that the ether is firmly fixed in space – that is to say, unable to move at all, and that electricity is firmly lodged in the mobile elementary particles. Today his discovery may be expressed as follows: Physical space and the ether are only different terms for the same thing; fields are physical conditions of space. For if no particular state of motion belongs to the ether, there does not seem to be any ground for introducing it as an entity of a special sort alongside of space. But the physicists were still far removed from such a way of thinking; space was still, for them, a rigid, homogeneous something, susceptible of no change or conditions. Only the genius of Riemann, solitary and uncomprehended, had already won its way by the middle of last century to a new conception of space, in which space was deprived of its rigidity, and in which its power to take part in physical events was recognized as possible. This intellectual achievement commands our admiration all the more for having preceded Faraday's and Clerk Maxwell's field theory of electricity. Then came the special theory of relativity with its recognition of the physical equivalence of all inertial systems. The inseparableness of time and space emerged in connection with electrodynamics, or the law of the propagation of light. Hitherto it had been silently assumed that the four-dimensional continuum of events could be split up into time and space in an objective manner – i.e., that an absolute significance attached to the 'now' in the world of events. With the discovery of the relativity of simultaneity, space and time were merged in a single continuum in[1] the same way as the three-dimensions of space had been before. Physical space was thus increased to a four-dimensional space which also included the dimension of time. The four-dimensional space of the special theory of relativity is just as rigid and absolute as Newton's space.

The theory of relativity is a fine example of the fun-

---

[1]EDITOR'S NOTE: The way this sentence is worded may leave readers, who are not acquainted with the historical facts – that it was Hermann Minkowski (Einstein's mathematics professor) who discovered the spacetime structure of the world – with the impression that it was Einstein who first demonstrated that "space and time were merged in a single continuum." And, indeed, even some physics students think that Einstein introduced the concept of spacetime in physics.

Such an impression is both historically inaccurate and unfair, especially given that

- it was Minkowski who (i) successfully decoded the message hidden in the unsuccessful *experiments* to detect absolute motion (that reality was a four-dimensional world), (ii) developed the four-dimensional mathematical formalism of flat spacetime physics, and (iii) demonstrated that what Einstein called special relativity was in fact a theory of an absolute four-dimensional world;

- Einstein's initial negative reaction to Minkowski's works [1]:

> Since the mathematicians have invaded the relativity theory, I do not understand it myself any more.

But later Einstein fully adopted the spacetime view of the world which was crucial for his general relativity.

1. Albert Einstein cited in: A. Sommerfeld, To Albert Einstein's Seventieth Birthday. In: *Albert Einstein: Philosopher-Scientist*. P. A. Schilpp, ed., 3rd ed. (Open Court, Illinois 1969) pp. 99-105, p. 102.

damental character of the modern development of theoretical science. The hypotheses with which it starts become steadily more abstract and remote from experience. On the other hand it gets nearer to the grand aim of all science, which is to cover the greatest possible number of empirical facts by logical deduction from the smallest possible number of hypotheses or axioms. Meanwhile the train of thought leading from the axioms to the empirical facts or verifiable consequences gets steadily longer and more subtle. The theoretical scientist is compelled in an increasing degree to be guided by purely mathematical, formal considerations in his search for a theory, because the physical experience of the experimenter cannot lift him into the regions of highest abstraction. The predominantly inductive methods appropriate to the youth of science are giving place to tentative deduction. Such a theoretical structure needs to be very thoroughly elaborated before it can lead to conclusions which can be compared with experience. Here too the observed fact is undoubtedly the supreme arbiter; but it cannot pronounce sentence until the wide chasm separating the axioms from their verifiable consequences has been bridged by much intense, hard thinking. The theorist has to set about this Herculean task in the clear consciousness that his efforts may only be destined to deal the death blow to his theory. The theorist who undertakes such a labor should not be carped at as "fanciful"; on the contrary, he should be encouraged to give free reign to his fancy, for there is no other way to the goal. His is no idle daydreaming, but a search for the logically simplest possibilities and their consequences. This plea was needed in order to make the hearer or reader more ready to follow the ensuing train of ideas with attention; it is the line of thought which has led from the special to the general theory of relativity and thence to its latest offshoot, the unitary field theory. In this exposition the use of mathematical symbols cannot be avoided.

We start with the special theory of relativity. This theory is still based directly on an empirical law, that of the constant velocity of light. Let P be a point in empty space, P' one separated from it by a length $d\sigma$ and infinitely near to it. Let a flash of light be emitted from P at a time t and reach P' at a time $t+dt$. Then

$$d\sigma^2 = C^2\, dt^2.$$

If dx_1, dx_2, dx_3 are the orthogonal projections of $d\sigma$, and the imaginary time coordinate $\sqrt{-ict} = x_4$ is introduced, then the above-mentioned law of the constancy of the propagation of light takes the form

$$ds^2 = dx_1^2 + dx_2^2 + dx_3^2 + dx_4^2 = 0.$$

Since this formula expresses a real situation, we may attribute a real meaning to the quantity ds, even supposing the neighboring points of the four-dimensional continuum are selected in such a way that the ds belonging to them does not disappear. This is more or less expressed by saying that the four-dimensional space (with imaginary time-coordinates) of the special theory of relativity possesses a Euclidean metric.

The fact that such a metric is called Euclidean is connected with the following. The position of such a metric in a three-dimensional continuum is fully equivalent to the positions of the axioms of Euclidean geometry. The defining equation of the metric is thus nothing but the Pythagorean theorem applied to the differentials and the coordinates.

Such alteration of the coordinates (by transformation) is permitted in the special theory of relativity, since in the new coordinates too the magnitude ds (fundamental invariant) is expressed in the new differentials of the coordinates by the sum of the squares. Such transformations are called Lorentz transformations.

The heuristic method of the special theory of relativity is characterized by the following principle:

Only those equations are admissible as an expression of natural laws which do not change their form when the coordinates are changed by means of a Lorentz transformation (covariance of equations in elation to Lorentz transformations).

This method led to the discovery of the necessary connection between impulse and energy, the strength of an electric and a magnetic field, electrostatic and electrodynamic forces, inert mass and energy; and the number of independent concepts and fundamental equations was thereby reduced.

This method pointed beyond itself. Is it true that the equations which express natural laws are covariant in relation to Lorentz transformations only and not in relation to other transformations? Well, formulated in that way the question really means nothing, since every system of equations can be expressed in general coordinates. We must ask. Are not the laws of nature so constituted that they receive no real simplification through the choice of any one particular set of coordinates?

We will only mention in passing that our empirical principle of the equality of inert and heavy masses prompts us to answer this question in the affirmative. If we elevate the equivalence of all coordinate systems for the formulation of natural laws into a principle, we arrive at the general theory of relativity, provided we stick to the law of the constant velocity of light or to the hypothesis of the objective significance of the Euclidean metric at least for infinitely small portions of four-dimensional space.

This means that for finite regions of space the existence (significant for physics) of a general Riemannian metric is presupposed according to the formula

$$ds^2 = \sum_{\mu\nu} g_{\mu\nu} dx^\mu dx^\nu,$$

whereby the summation is to be extended to all index combinations from 11 to 44.

The structure of such a space differs absolutely radically in one respect from that of a Euclidean space. The coefficients $g_{\mu\nu}$ are for the time being any functions whatever of the coordinates x_1 to x_4, and the structure of the space is not really determined until these functions $g_{\mu\nu}$ are really known. It is only determined more closely by specifying laws which the metrical field of the $g_{\mu\nu}$ satisfy. On physical grounds this gave rise to the conviction that the metrical field was at the same time the gravitational field.

Since the gravitational field is determined by the configuration of masses and changes with it, the geometric structure of this space is also dependent on physical factors. Thus according to this theory space is – exactly as Riemann guessed – no longer absolute; its structure depends on physical influences. Physical geometry is no longer an isolated self-contained science like the geometry of Euclid.

The problem of gravitation was thus reduced to a mathematical problem: it was required to find the simplest fundamental equations which are covariant in relation to any transformation of coordinates whatever.

I will not speak here of the way this theory has been confirmed by experience, but explain at once why Theory could not rest permanently satisfied with this success. Gravitation had indeed been traced to the structure of space, but besides the gravitational field there is also the electromagnetic field. This had, to begin with, to be introduced into the theory as an entity independent of gravitation. Additional terms which took account of the existence of the electromagnetic field had to be included in the fundamental equations for the field. But the idea that there were two structures of space independent of each other, the metric-gravitational and the electromagnetic, was intolerable to the theoretical spirit. We are forced to the belief that both sorts of field must correspond to verified structure of space.

The "unitary field-theory," which represents itself as a mathematically independent extension of the general theory of relativity, attempts to fulfill this last postulate of the field theory. The formal problem should be put as follows: Is there a theory of the continuum in which a new structural element ap- pears side by side with the metric such that it forms a single whole together with the metric? If so, what are the simplest field laws to which such a continuum can be made subject? And finally, are these field-laws well fitted to represent the properties of the gravitational field and the electromagnetic field? Then there is the further question whether the corpuscles (electrons and protons) can be regarded as positions of particularly dense fields, whose movements are determined by the field equations. At present there is only one way of answering the first three questions. The space structure on which it is based may be described as follows, and the description applies equally to a space of any number of dimensions.

Space has a Riemannian metric. This means that the Euclidean geometry holds good in the infinitesimal neighborhood of every point P. Thus for the neighborhood of every point P there is a local Cartesian system of coordinates, in reference to which the metric is calculated according to the Pythagorean theorem. If we now imagine the length 1 cut off from the positive axes of these local systems, we get the orthogonal "local n-leg." Such a local n-leg is to be found in every other point P' of space also. Thus, if a linear element (PG or $P'G'$) starting from the points P or P', is given, then the magnitude of this linear element can be calculated by the aid of the relevant local n-leg from its local coordinates by means of Pythagoras's theorem. There is therefore a definite meaning in speaking of the numerical equality of the linear elements PG and $P'G'$.

It is essential to observe now that the local orthogonal n-legs are not completely determined by the metric. For we can still select the orientation of the n-legs perfectly freely without causing any alteration in the result of calculating the size of the linear elements according to Pythagoras's theorem. A corollary of this is that in a space whose structure consists exclusively of a Riemannian metric, two linear elements PG and $P'G'$, can be compared with regard to their magnitude but not their direction; in particular, there is no sort of point in saying that the two linear elements are parallel to one another. In this respect, therefore, the purely metrical (Riemannian) space is less rich in structure than the Euclidean.

Since we are looking for a space which exceeds Riemannian space in wealth of structure, the obvious thing is to enrich Riemannian space by adding the relation of direction or parallelism. Therefore for every direction through P let there be a definite direction through P', and let this mutual relation be a determinate one. We call the directions thus related to each other "parallel." Let this parallel relation further fulfill the condition of angular uniformity: If PG and PK are two directions in P, $P'G'$ and $P'K'$ the corresponding parallel directions through P', then the angles KPG and $K'P'G'$ (measurable on Euclidean lines in the local system) should be equal.

The basic space-structure is thereby completely defined. It is most easily described mathematically as follows: In the definite point P we suppose an orthogonal n-leg with definite, freely chosen orientation. In every other point P' of space we so orient its local n-leg that its axes are parallel to the corresponding axes at the point P. Given the above structure of space and free choice in the orientation of the n-leg at one point P, all n-legs are thereby completely defined. In the space P let us now imagine any Gaussian system of coordinates and that in every point the axes of the n-leg there are projected on to it. This system of n components completely describes the structure of space.

This spatial structure stands, in a sense, midway between the Riemannian and the Euclidean. In contrast to the former, it has room for the straight-line, that is to say a line all of whose elements are parallel to each other in pairs. The geometry here described differs from the Euclidean in the non-existence of the parallelogram. If at the ends P and G of a length PG two equal and parallel lengths PP' and GG' are marked off, $P'G'$ is in general neither equal nor parallel to PG.

The mathematical problem now solved so far is

this: What are the simplest conditions to which a space-structure of the kind described can be subjected? The chief question which still remains to be investigated is this: To what extent can physical fields and primary entities be represented by solutions, free from singularities, of the equations which answer the former question?

SIDELIGHTS ON RELATIVITY*

BY

ALBERT EINSTEIN

Translated by
G. B. Jeffery and W. Perrett

---

*A. Einstein, *Sidelights on Relativity*. Translated by G. B. Jeffery and W. Perrett (Methuen & Co. ltd., London 1922).

ETHER AND THE THEORY OF RELATIVITY

An Address delivered on May 5th, 1920 in the University of Leyden

How does it come about that alongside of the idea of ponderable matter, which is derived by abstraction from everyday life, the physicists set the idea of the existence of another kind of matter, the ether? The explanation is probably to be sought in those phenomena which have given rise to the theory of action at a distance, and in the properties of light which have led to the undulatory theory. Let us devote a little while to the consideration of these two subjects.

Outside of physics we know nothing of action at a distance. When we try to connect cause and effect in the experiences which natural objects afford us, it seems at first as if there were no other mutual actions than those of immediate contact, e.g. the communication of motion by impact, push and pull, heating or inducing combustion by means of a flame, etc. It is true that even in everyday experience weight, which is in a sense action at a distance, plays a very important part. But since in daily experience the weight of bodies meets us as something constant, something not linked to any cause which is variable in time or place, we do not in everyday life speculate as to the cause of gravity, and therefore do not become conscious of its character as action at a distance. It was Newton's theory of gravitation that first assigned a cause for gravity by interpreting it as action at a distance, proceeding from masses. Newton's theory is probably the greatest stride ever made in the effort towards the causal nexus of natural phenomena. And yet this theory evoked a lively sense of discomfort among Newton's contemporaries, because it seemed to be in conflict with the principle springing from the rest of experience, that there can be reciprocal action only through contact, and not through immediate action at a distance.

It is only with reluctance that man's desire for knowledge endures a dualism of this kind. How was unity to be preserved in his comprehension of the forces of nature? Either by trying to look upon contact forces as being themselves distant forces which admittedly are observable only at a very small distance – and this was the road which Newton's followers, who were entirely under the spell of his doctrine, mostly preferred to take; or by assuming that the Newtonian action at a distance is only *apparently* immediate action at a distance, but in truth is conveyed by a medium permeating space, whether by movements or by elastic deformation of this medium. Thus the endeavour toward a unified view of the nature of forces leads to the hypothesis of an ether. This hypothesis, to be sure, did not at first bring with it any advance in the theory of gravitation or in physics generally, so that it became customary to treat Newton's law of force as an axiom not further reducible. But the ether hypothesis was bound always to play some part in physical science, even if at first only a latent part.

When in the first half of the nineteenth century the far-reaching similarity was revealed which subsists between the properties of light and those of elastic waves in ponderable bodies, the ether hypothesis found fresh support. It appeared beyond question that light must be interpreted as a vibratory process in an elastic, inert medium filling up universal space. It also seemed to be a necessary consequence of the fact that light is capable of polarisation that this medium, the ether, must be of the nature of a solid body, because transverse waves are not possible in a fluid, but only in a solid. Thus the physicists were bound to arrive at the theory of the "quasi-rigid" luminiferous ether, the parts of which can carry out no movements relatively to one another except the small movements of deformation which correspond to light-waves.

This theory – also called the theory of the stationary luminiferous ether – moreover found a strong support in an experiment which is also of fundamental importance in the special theory of relativity, the experiment of Fizeau, from which one was obliged to infer that the luminiferous ether does not take part in the movements of bodies. The phenomenon of aberration also favoured the theory of the quasi-rigid ether.

The development of the theory of electricity along the path opened up by Maxwell and Lorentz gave the development of our ideas concerning the ether quite a peculiar and unexpected turn. For Maxwell himself the ether indeed still had properties which were purely mechanical, although of a much more complicated kind than the mechanical properties of tangible solid bodies. But neither Maxwell nor his followers succeeded in elaborating a mechanical model for the ether which might furnish a satisfactory mechanical interpretation of Maxwell's laws of the electromagnetic field. The laws were clear and simple, the mechanical interpretations clumsy and contradictory. Almost imperceptibly the theoretical physicists adapted themselves to a situation which, from the standpoint of their mechanical programme, was very depressing. They were particularly influenced by the electrodynamical investigations of Heinrich Hertz. For whereas they previously had required of a conclusive theory that it should content itself with the fundamental concepts which belong exclusively to mechanics (e.g. densities, velocities, deformations, stresses) they gradually accustomed themselves to admitting electric and magnetic force as fundamental concepts side by side with those of mechanics, without requiring a mechanical interpretation for them. Thus the purely mechanical view of nature was gradually abandoned. But this change led to a fundamental dualism which in the long-run was insupportable. A way of escape was now sought in the reverse direction, by reducing the principles of mechanics to those of electricity, and this especially as confidence in the strict validity of the equations of Newton's mechanics was shaken by the experiments with β-rays and rapid kathode rays.

This dualism still confronts us in unextenuated form in the theory of Hertz, where matter appears not only as the bearer of velocities, kinetic energy, and mechanical pressures, but also as the bearer of electromagnetic fields. Since such fields also occur *in vacuo* – i.e. in free ether – the ether also appears as bearer of electromagnetic fields. The ether appears indistinguishable in its functions from ordinary matter. Within matter it takes part in the motion of matter and in empty space it has everywhere a velocity; so that the ether has a definitely assigned velocity throughout the whole of space. There is no fundamental difference between Hertz's ether and ponderable matter (which in part subsists in the ether).

The Hertz theory suffered not only from the defect of ascribing to matter and ether, on the one hand mechanical states, and on the other hand electrical states, which do not stand in any conceivable relation to each other; it was also at variance with the result of Fizeau's important experiment on the velocity of the propagation of light in moving fluids, and with other established experimental results.

Such was the state of things when H. A. Lorentz entered upon the scene. He brought theory into harmony with experience by means of a wonderful simplification of theoretical principles. He achieved this, the most important advance in the theory of electricity since Maxwell, by taking from ether its mechanical, and from matter its electromagnetic qualities. As in empty space, so too in the interior of material bodies, the ether, and not matter viewed atomistically, was exclusively the seat of electromagnetic fields. According to Lorentz the elementary particles of matter alone are capable of carrying out movements; their electromagnetic activity is entirely confined to the carrying of electric charges. Thus Lorentz succeeded in reducing all electromagnetic happenings to Maxwell's equations for free space.

As to the mechanical nature of the Lorentzian ether, it may be said of it, in a somewhat playful spirit, that immobility is the only mechanical property of which it has not been deprived by H. A. Lorentz. It may be added that the whole change in the conception of the ether which the special theory of relativity brought about, consisted in taking away from the ether its last mechanical quality, namely, its immobility. How this is to be understood will forthwith be expounded.

The spacetime theory and the kinematics of the special theory of relativity were modelled on the

Maxwell-Lorentz theory of the electromagnetic field. This theory therefore satisfies the conditions of the special theory of relativity, but when viewed from the latter it acquires a novel aspect. For if K be a system of coordinates relatively to which the Lorentzian ether is at rest, the Maxwell-Lorentz equations are valid primarily with reference to K. But by the special theory of relativity the same equations without any change of meaning also hold in relation to any new system of coordinates K' which is moving in uniform translation relatively to K. Now comes the anxious question: Why must I in the theory distinguish the K system above all K' systems, which are physically equivalent to it in all respects, by assuming that the ether is at rest relatively to the K system? For the theoretician such an asymmetry in the theoretical structure, with no corresponding asymmetry in the system of experience, is intolerable. If we assume the ether to be at rest relatively to K, but in motion relatively to K', the physical equivalence of K and K' seems to me from the logical standpoint, not indeed downright incorrect, but nevertheless unacceptable.

The next position which it was possible to take up in face of this state of things appeared to be the following. The ether does not exist at all. The electromagnetic fields are not states of a medium, and are not bound down to any bearer, but they are independent realities which are not reducible to anything else, exactly like the atoms of ponderable matter. This conception suggests itself the more readily as, according to Lorentz's theory, electromagnetic radiation, like ponderable matter, brings impulse and energy with it, and as, according to the special theory of relativity, both matter and radiation are but special forms of distributed energy, ponderable mass losing its isolation and appearing as a special form of energy.

More careful reflection teaches us, however, that the special theory of relativity does not compel us to deny ether. We may assume the existence of an ether; only we must give up ascribing a definite state of motion to it, i.e. we must by abstraction take from it the last mechanical characteristic which Lorentz had still left it. We shall see later that this point of view, the conceivability of which I shall at once endeavour to make more intelligible by a somewhat halting comparison, is justified by the results of the general theory of relativity.

Think of waves on the surface of water. Here we can describe two entirely different things. Either we may observe how the undulatory surface forming the boundary between water and air alters in the course of time; or else – with the help of small floats, for instance – we can observe how the position of the separate particles of water alters in the course of time. If the existence of such floats for tracking the motion of the particles of a fluid were a fundamental impossibility in physics – if, in fact, nothing else whatever were observable than the shape of the space occupied by the water as it varies in time, we should have no ground for the assumption that water consists of movable particles. But all the same we could characterise it as a medium.

We have something like this in the electromagnetic field. For we may picture the field to ourselves as consisting of lines of force. If we wish to interpret these lines of force to ourselves as something material in the ordinary sense, we are tempted to interpret the dynamic processes as motions of these lines of force, such that each separate line of force is tracked through the course of time. It is well known, however, that this way of regarding the electromagnetic field leads to contradictions.

Generalising we must say this: There may be supposed to be extended physical objects to which the idea of motion cannot be applied. They may not be thought of as consisting of particles which allow themselves to be separately tracked through time. In Minkowski's idiom this is expressed as follows: -Not every extended conformation in the four-dimensional world can be regarded as composed of world-threads. The special theory of relativity forbids us to assume the ether to consist of particles observable through time, but the hypothesis of ether in itself is not in conflict with the special theory of relativity. Only we must be on our guard against ascribing a state of motion to the ether.

Certainly, from the standpoint of the special theory of relativity, the ether hypothesis appears at first to be an empty hypothesis. In the equations of the electromagnetic field there occur, in addition to the densities of the electric charge, *only* the intensities of the field. The career of electromagnetic processes *in vacuo* appears to be completely determined by these equations, uninfluenced by other physical quantities. The electromagnetic fields appear as ultimate, irreducible realities, and at first it seems superfluous to postulate a homogeneous, isotropic ether-medium, and to envisage electromagnetic fields as states of this medium.

But on the other hand there is a weighty argument to be adduced in favour of the ether hypothesis. To deny the ether is ultimately to assume that empty space has no physical qualities whatever. The fundamental facts of mechanics do not harmonize with this view. For the mechanical behaviour of a corporeal system hovering freely in empty space depends not only on relative positions (distances) and relative velocities, but also on its state of rotation, which physically may be taken as a characteristic not appertaining to the system in itself. In order to be able to look upon the rotation of the system, at least formally, as something real, Newton objectivises space. Since he classes his absolute space together with real things, for him rotation relative to an absolute space is also something real. Newton might no less well have called his absolute space "Ether"; what is essential is merely that besides observable objects, another thing, which is not perceptible, must be looked upon as real, to enable acceleration or rotation to be looked upon as something real.

It is true that Mach tried to avoid having to accept as real something which is not observable by endeavouring to substitute in mechanics a mean acceleration with reference to the totality of the masses in the universe in place of an acceleration with reference to absolute space. But inertial resistance opposed to relative acceleration of distant masses presupposes action at a distance; and as the modern physicist does not believe that he may accept this action at a distance, he comes back once more, if he follows Mach, to the ether, which has to serve as medium for the effects of inertia. But this conception of the ether to which we are led by Mach's way of thinking differs essentially from the ether as conceived by Newton, by Fresnel, and by Lorentz. Mach's ether not only *conditions* the behaviour of inert masses, but is also *conditioned* in its state by them.

Mach's idea finds its full development in the ether of the general theory of relativity. According to this theory the metrical qualities of the continuum of space-time differ in the environment of different points of spacetime, and are partly *conditioned* by the matter existing outside of the territory under consideration. This spacetime variability of the reciprocal relations of the standards of space and time, or, perhaps, the recognition of the fact that "empty space" in its physical relation is neither homogeneous nor isotropic, compelling us to describe its state by ten functions (the gravitation potentials $g_{\mu\nu}$), has, I think, finally disposed of the view that space is physically empty. But therewith the conception of the ether has again acquired an intelligible content, although this content differs widely from that of the ether of the mechanical undulatory theory of light. The ether of the general theory of relativity is a medium which is itself devoid of *all* mechanical and kinematical qualities, but helps to determine mechanical (and electromagnetic) events.

What is fundamentally new in the ether of the general theory of relativity as opposed to the ether of Lorentz consists in this, that the state of the former is at every place determined by connections with the matter and the state of the ether in neighbouring places, which are amenable to law in the form of differential equations; whereas the state of the Lorentzian ether in the absence of electromagnetic fields is conditioned by nothing outside itself, and is everywhere the same. The ether of the general theory of relativity is transmuted conceptually into the ether of Lorentz if we substitute constants for the functions of space which describe the former, disregarding the causes which condition its state. Thus we may also say, I think, that the ether of the general theory of relativity is the outcome of the Lorentzian ether, through relativation.

As to the part which the new ether is to play in the physics of the future we are not yet clear. We know that it determines the metrical relations in the space-time continuum, e.g. the configurative possibilities of solid bodies as well as the gravitational fields; but we do not know whether it has an essential share in the structure of the electrical elementary particles constituting matter. Nor do we know whether it is only in the proximity of ponderable masses that its structure differs essentially from that of the Lorentzian ether; whether the geometry of spaces of cosmic extent is approximately Euclidean. But we can assert by reason of the relativistic equations of gravitation that there must be a departure from Euclidean relations, with spaces of cosmic order of magnitude, if there exists a positive mean density, no matter how small, of the matter in the universe. In this case the universe must of necessity be spatially unbounded and of finite magnitude, its magnitude being determined by the value of that mean density.

If we consider the gravitational field and the electromagnetic field from the stand-point of the ether hypothesis, we find a remarkable difference between the two. There can be no space nor any part of space without gravitational potentials; for these confer upon space its metrical qualities, without which it cannot be imagined at all. The existence of the gravitational field is inseparably bound up with the existence of space. On the other hand a part of space may very well be imagined without an electromagnetic field; thus in contrast with the gravitational field, the electromagnetic field seems to be only secondarily linked to the ether, the formal nature of the electromagnetic field being as yet in no way determined by that of gravitational ether. From the present state of theory it looks as if the electromagnetic field, as opposed to the gravitational field, rests upon an entirely new formal *motif,* as though nature might just as well have endowed the gravitational ether with fields of quite another type, for example, with fields of a scalar potential, instead of fields of the electromagnetic type.

Since according to our present conceptions the elementary particles of matter are also, in their essence, nothing else than condensations of the electromagnetic field, our present view of the universe presents two realities which are completely separated from each other conceptually, although connected causally, namely, gravitational ether and electromagnetic field, or – as they might also be called – space and matter.

Of course it would be a great advance if we could succeed in comprehending the gravitational field and the electromagnetic field together as one unified conformation. Then for the first time the epoch of theoretical physics founded by Faraday and Maxwell would reach a satisfactory conclusion. The contrast between ether and matter would fade away, and, through the general theory of relativity, the whole of physics would become a complete system of thought, like geometry, kinematics, and the theory of gravitation. An exceedingly ingenious attempt in this direction has been made by the mathematician H. Weyl; but I do not believe that his theory will hold its ground in relation to reality. Further, in contemplating the immediate future of theoretical physics we ought not unconditionally to reject the possibility that the facts comprised in the quantum theory may set bounds to the field theory beyond which it cannot pass.

Recapitulating, we may say that according to the general theory of relativity space is endowed with physical qualities; in this sense, therefore, there exists an ether. According to the general theory of relativity space without ether is unthinkable; for in such space there not only would be no propagation of light, but also no possibility of existence for standards of space and time (measuring-rods and clocks), nor therefore any spacetime intervals in the physical sense.[1] But

---

[1]EDITOR'S NOTE: To reach this conclusion, it took Einstein 15 years – from the publication of his paper on special relativity (1905) to 1920 when he gave this lecture in the University of Leyden. In 1905 he wrote [1]:

> The introduction of a "luminiferous ether" will prove to be superfluous inasmuch as the view here to be developed will not require an "absolutely stationary space" provided with special properties, nor assign a velocity-vector to a point of the empty space in which electromagnetic processes take place.

So from stating that the ether has no place in special relativity to explaining in this paragraph that:

> According to the general theory of relativity space without ether is unthinkable; for in such space there not only would be no propagation of light, but also no possibility of existence for standards of space and time (measuring-rods and clocks), nor therefore any spacetime intervals in the physical sense.

It looks reasonably certain that it took so long even to Einstein – the creator of the greatest intellectual achievement in fundamental physics (general relativity) – to arrive at the view on the ether presented in this lecture, because he had been significantly influenced by Mach's ideas on the nature of motion and space. In fact, the arguments in the above quote fully apply in the case of flat spacetime (where special relativity is valid). This can be clearly seen in Minkowski's 1908 lecture "Space and Time" [2] and also from Dirac's position on ether expressed in a Letter to the Editor of *Nature* [3]:

> Since it is the medium the condition of which determined light and electromagnetic force, we may call it *aether*; since it is the subject-matter of the science of geometry, we may call it *space*; some-

this ether may not be thought of as endowed with the quality characteristic of ponderable media, as consisting of parts which may be tracked through time. The idea of motion may not be applied to it.

---

times in order to avoid giving preference to either aspect, it is called by Minkowski's term *world*.

In his book *The Einstein Theory of Relativity: A Concise Statement*, included in this collection, Lorentz also pointed out that "It is not necessary to give up entirely even the ether," p. 102.

1. A. Einstein, "On the Electrodynamics of Moving Bodies" in *The Origin of Spacetime Physics*, With a Foreword by Abhay Ashtekar. Edited by Vesselin Petkov (Minkowski Institute Press, Montreal 2020), p. 85.

2. H. Minkowski, "Space and Time" in: H. Minkowski, *Spacetime: Minkowski's Papers on Spacetime Physics*. Edited with an Introduction by Vesselin Petkov (Minkowski Institute Press, Montreal 2020), pp. 57-76, p. 62.

3. A. S. Eddington, "Space" or "Aether"? *Nature* **107**, 201 (April 14, 1921). It is reprinted in: A. S. Eddington, *Space, Time and Gravitation: An Outline of the General Relativity Theory*. With a Foreword by Dennis Dieks. Edited by Vesselin Petkov (Minkowski Institute Press, Montreal 2017).

Geometry and Experience

An expanded form of an Address to the Prussian Academy of Sciences in Berlin on January 27th, 1921

One reason why mathematics enjoys special esteem, above all other sciences, is that its laws are absolutely certain and indisputable, while those of all other sciences are to some extent debatable and in constant danger of being overthrown by newly discovered facts. In spite of this, the investigator in another department of science would not need to envy the mathematician if the laws of mathematics referred to objects of our mere imagination, and not to objects of reality. For it cannot occasion surprise that different persons should arrive at the same logical conclusions when they have already agreed upon the fundamental laws (axioms), as well as the methods by which other laws are to be deduced therefrom. But there is another reason for the high repute of mathematics, in that it is mathematics which affords the exact natural sciences a certain measure of security, to which without mathematics they could not attain.

At this point an enigma presents itself which in all ages has agitated inquiring minds. How can it be that mathematics, being after all a product of human thought which is independent of experience, is so admirably appropriate to the objects of reality? Is human reason, then, without experience, merely by taking thought, able to fathom the properties of real things.

In my opinion the answer to this question is, briefly, this: As far as the laws of mathematics refer to reality, they are not certain; and as far as they are certain, they do not refer to reality. It seems to me that complete clearness as to this state of things first became common property through that new departure in mathematics which is known by the name of mathematical logic or "Axiomatics." The progress achieved by axiomatics consists in its having neatly separated the logical-formal from its objective or intuitive content; according to axiomatics the logical-formal alone forms the subject-matter of mathematics, which is not concerned with the intuitive or other content associated with the logical-formal.

Let us for a moment consider from this point of view any axiom of geometry, for instance, the following: Through two points in space there always passes one and only one straight line. How is this axiom to be interpreted in the older sense and in the more modern sense?

The older interpretation: Every one knows what a straight line is, and what a point is. Whether this knowledge springs from an ability of the human mind or from experience, from some collaboration of the two or from some other source, is not for the mathematician to decide. He leaves the question to the philosopher. Being based upon this knowledge, which precedes all mathematics, the axiom stated above is, like all other axioms, self-evident, that is, it is the expression of a part of this *à priori* knowledge.

The more modern interpretation: Geometry treats

of entities which are denoted by the words straight line, point, etc. These entities do not take for granted any knowledge or intuition whatever, but they presuppose only the validity of the axioms, such as the one stated above, which are to be taken in a purely formal sense, i.e. as void of all content of intuition or experience. These axioms are free creations of the human mind. All other propositions of geometry are logical inferences from the axioms (which are to be taken in the nominalistic sense only). The matter of which geometry treats is first defined by the axioms. Schlick in his book on epistemology has therefore characterised axioms very aptly as "implicit definitions."

This view of axioms, advocated by modern axiomatics, purges mathematics of all extraneous elements, and thus dispels the mystic obscurity which formerly surrounded the principles of mathematics.

But a presentation of its principles thus clarified makes it also evident that mathematics as such cannot predicate anything about perceptual objects or real objects. In axiomatic geometry the words "point," "straight line," etc., stand only for empty conceptual schemata. That which gives them substance is not relevant to mathematics.

Yet on the other hand it is certain that mathematics generally, and particularly geometry, owes its existence to the need which was felt of learning something about the relations of real things to one another. The very word geometry, which, of course, means earth-measuring, proves this. For earth-measuring has to do with the possibilities of the disposition of certain natural objects with respect to one another, namely, with parts of the earth, measuring-lines, measuring-wands, etc. It is clear that the system of concepts of axiomatic geometry alone cannot make any assertions as to the relations of real objects of this kind, which we will call practically-rigid bodies. To be able to make such assertions, geometry must be stripped of its merely logical-formal character by the co-ordination of real objects of experience with the empty conceptual frame-work of axiomatic geometry. To accomplish this, we need only add the proposition: Solid bodies are related, with respect to their possible dispositions, as are bodies in Euclidean geometry of three dimensions. Then the propositions of Euclid contain affirmations as to the relations of practically-rigid bodies.

Geometry thus completed is evidently a natural science; we may in fact regard it as the most ancient branch of physics. Its affirmations rest essentially on induction from experience, but not on logical inferences only. We will call this completed geometry "practical geometry," and shall distinguish it in what follows from "purely axiomatic geometry." The question whether the practical geometry of the universe is Euclidean or not has a clear meaning, and its answer can only be furnished by experience. All linear measurement in physics is practical geometry in this sense, so too is geodetic and astronomical linear measurement, if we call to our help the law of experience that light is propagated in a straight line, and indeed in a straight line in the sense of practical geometry.

I attach special importance to the view of geometry which I have just set forth, because without it I should have been unable to formulate the theory of relativity. Without it the following reflection would have been impossible: In a system of reference rotating relatively to an inert system, the laws of disposition of rigid bodies do not correspond to the rules of Euclidean geometry on account of the Lorentz contraction; thus if we admit non-inert systems we must abandon Euclidean geometry. The decisive step in the transition to general co-variant equations would certainly not have been taken if the above interpretation had not served as a stepping-stone. If we deny the relation between the body of axiomatic Euclidean geometry and the practically-rigid body of reality, we readily arrive at the following view, which was entertained by that acute and profound thinker, H. Poincaré: Euclidean geometry is distinguished above all other imaginable axiomatic geometries by its simplicity. Now since axiomatic geometry by itself contains no assertions as to the reality which can be experienced, but can do so only in combination with physical laws, it should be possible and reasonable – whatever may be the nature of reality – to retain Euclidean geometry. For if contradictions between theory and experience manifest themselves, we should rather decide to change physical laws than to change axiomatic Euclidean geometry. If we deny the relation between the practically-rigid body and geometry, we shall indeed not easily free ourselves from the convention that Euclidean geometry is to be retained as the simplest. Why is the equivalence of the practically-rigid body and the body of geometry – which suggests itself so readily – denied by Poincaré and other investigators? Simply because under closer inspection the real solid bodies in nature are not rigid, because their geometrical behaviour, that is, their possibilities of relative disposition, depend upon temperature, external forces, etc. Thus the original, immediate relation between geometry and physical reality appears destroyed, and we feel impelled toward the following more general view, which characterizes Poincaré's standpoint. Geometry (G) predicates nothing about the relations of real things, but only geometry together with the purport (P) of physical laws can do so. Using symbols, we may say that only the sum of (G) + (P) is subject to the control of experience. Thus (G) may be chosen arbitrarily, and also parts of (P); all these laws are conventions. All that is necessary to avoid contradictions is to choose the remainder of (P) so that (G) and the whole of (P) are together in accord with experience. Envisaged in this way, axiomatic geometry and the part of natural law which has been given a conventional status appear as epistemologically equivalent.

Sub specie aeterni Poincaré, in my opinion, is right. The idea of the measuring-rod and the idea of the clock coordinated with it in the theory of relativity do not find their exact correspondence in the real world. It is also clear that the solid body and the clock do not in the conceptual edifice of physics play the part of irreducible elements, but that of composite structures, which may not play any independent part in theoretical physics. But it is my conviction that in the present stage of development of theoretical physics these ideas must still be employed as independent ideas; for we are still far from possessing such certain knowledge of theoretical principles as to be able to give exact theoretical constructions of solid bodies and clocks.

Further, as to the objection that there are no really rigid bodies in nature, and that therefore the properties predicated of rigid bodies do not apply to physical reality, – this objection is by no means so radical as might appear from a hasty examination. For it is not a difficult task to determine the physical state of a measuring-rod so accurately that its behaviour relatively to other measuring-bodies shall be sufficiently free from ambiguity to allow it to be substituted for the "rigid" body. It is to measuring-bodies of this kind that statements as to rigid bodies must be referred.

All practical geometry is based upon a principle which is accessible to experience, and which we will now try to realise. We will call that which is enclosed between two boundaries, marked upon a practically-rigid body, a tract. We imagine two practically-rigid bodies, each with a tract marked out on it. These two tracts are said to be "equal to one another" if the boundaries of the one tract can be brought to coincide permanently with the boundaries of the other. We now assume that:

If two tracts are found to be equal once and anywhere, they are equal always and everywhere.

Not only the practical geometry of Euclid, but also its nearest generalisation, the practical geometry of Riemann, and therewith the general theory of relativity, rest upon this assumption. Of the experimental reasons which warrant this assumption I will mention only one. The phenomenon of the propagation of light in empty space assigns a tract, namely, the appropriate path of light, to each interval of local time, and conversely. Thence it follows that the above assumption for tracts must also hold good for intervals of clock-time in the theory of relativity. Consequently it may be formulated as follows: If two ideal clocks are going at the same rate at any time and at any place (being then in immediate proximity to each other), they will always go at the same rate, no matter where and when they are again compared with each other at one place. – If this law were not valid for real clocks, the proper frequencies for the separate atoms of the same chemical element would not be in such exact agreement as experience demonstrates. The existence of sharp spectral lines is a convincing experimental proof of the above-mentioned principle of practical geometry. This is the ultimate foundation in fact which enables us to speak with meaning of the mensuration, in Riemann's sense of the word, of the four-dimensional continuum of spacetime.

The question whether the structure of this continuum is Euclidean, or in accordance with Riemann's general scheme, or otherwise, is, according to the view which is here being advocated, properly speaking a physical question which must be answered by experience, and not a question of a mere convention to be selected on practical grounds. Riemann's geometry will be the right thing if the laws of disposition of practically-rigid bodies are transformable into those of the bodies of Euclid's geometry with an exactitude which increases in proportion as the dimensions of the part of spacetime under consideration are diminished.

It is true that this proposed physical interpretation of geometry breaks down when applied immediately to spaces of sub-molecular order of magnitude. But nevertheless, even in questions as to the constitution of elementary particles, it retains part of its importance. For even when it is a question of describing the electrical elementary particles constituting matter, the attempt may still be made to ascribe physical importance to those ideas of fields which have been physically defined for the purpose of describing the geometrical behaviour of bodies which are large as compared with the molecule. Success alone can decide as to the justification of such an attempt, which postulates physical reality for the fundamental principles of Riemann's geometry outside of the domain of their physical definitions. It might possibly turn out that this extrapolation has no better warrant than the extrapolation of the idea of temperature to parts of a body of molecular order of magnitude.

It appears less problematical to extend the ideas of

practical geometry to spaces of cosmic order of magnitude. It might, of course, be objected that a construction composed of solid rods departs more and more from ideal rigidity in proportion as its spatial extent becomes greater. But it will hardly be possible, I think, to assign fundamental significance to this objection. Therefore the question whether the universe is spatially finite or not seems to me decidedly a pregnant question in the sense of practical geometry. I do not even consider it impossible that this question will be answered before long by astronomy. Let us call to mind what the general theory of relativity teaches in this respect. It offers two possibilities:

1. The universe is spatially infinite. This can be so only if the average spatial density of the matter in universal space, concentrated in the stars, vanishes, i.e. if the ratio of the total mass of the stars to the magnitude of the space through which they are scattered approximates indefinitely to the value zero when the spaces taken into consideration are constantly greater and greater.

2. The universe is spatially finite. This must be so, if there is a mean density of the ponderable matter in universal space differing from zero. The smaller that mean density, the greater is the volume of universal space.

I must not fail to mention that a theoretical argument can be adduced in favour of the hypothesis of a finite universe. The general theory of relativity teaches that the inertia of a given body is greater as there are more ponderable masses in proximity to it; thus it seems very natural to reduce the total effect of inertia of a body to action and reaction between it and the other bodies in the universe, as indeed, ever since Newton's time, gravity has been completely reduced to action and reaction between bodies. From the equations of the general theory of relativity it can be deduced that this total reduction of inertia to reciprocal action between masses – as required by E. Mach, for example – is possible only if the universe is spatially finite.

On many physicists and astronomers this argument makes no impression. Experience alone can finally decide which of the two possibilities is realised in nature. How can experience furnish an answer? At first it might seem possible to determine the mean density of matter by observation of that part of the universe which is accessible to our perception. This hope is illusory. The distribution of the visible stars is extremely irregular, so that we on no account may venture to set down the mean density of star-matter in the universe as equal, let us say, to the mean density in the Milky Way. In any case, however great the space examined may be, we could not feel convinced that there were no more stars beyond that space. So it seems impossible to estimate the mean density. But there is another road, which seems to me more practicable, although it also presents great difficulties. For if we inquire into the deviations shown by the consequences of the general theory of relativity which are accessible to experience, when these are compared with the consequences of the Newtonian theory, we first of all find a deviation which shows itself in close proximity to gravitating mass, and has been confirmed in the case of the planet Mercury. But if the universe is spatially finite there is a second deviation from the Newtonian theory, which, in the language of the Newtonian theory, may be expressed thus: The gravitational field is in its nature such as if it were produced, not only by the ponderable masses, but also by a mass-density of negative sign, distributed uniformly throughout space. Since this factitious mass-density would have to be enormously small, it could make its presence felt only in gravitating systems of very great extent.

Assuming that we know, let us say, the statistical distribution of the stars in the Milky Way, as well as their masses, then by Newton's law we can calculate the gravitational field and the mean velocities which the stars must have, so that the Milky Way should not collapse under the mutual attraction of its stars, but should maintain its actual extent. Now if the actual velocities of the stars, which can, of course, be measured, were smaller than the calculated velocities, we should have a proof that the actual attractions at great distances are smaller than by Newton's law. From such a deviation it could be proved indirectly that the universe is finite. It would even be possible to estimate its spatial magnitude.

Can we picture to ourselves a three-dimensional universe which is finite, yet unbounded?

The usual answer to this question is "No," but that is not the right answer. The purpose of the following remarks is to show that the answer should be "Yes." I want to show that without any extraordinary difficulty we can illustrate the theory of a finite universe by means of a mental image to which, with some practice, we shall soon grow accustomed.

First of all, an observation of epistemological nature. A geometrical-physical theory as such is incapable of being directly pictured, being merely a system of concepts. But these concepts serve the purpose of bringing a multiplicity of real or imaginary sensory experiences into connection in the mind. To "visualise" a theory, or bring it home to one's mind, therefore means to give a representation to that abundance of experiences for which the theory supplies the schematic arrangement. In the present case we have to ask ourselves how we can represent that relation of solid bodies with respect to their reciprocal disposition (contact) which corresponds to the theory of a finite universe. There is really nothing new in what I have to say about this; but innumerable questions addressed to me prove that the requirements of those who thirst for knowledge of these matters have not yet been completely satisfied.

So, will the initiated please pardon me, if part of what I shall bring forward has long been known?

What do we wish to express when we say that our space is infinite? Nothing more than that we might lay any number whatever of bodies of equal sizes side by side without ever filling space. Suppose that we are provided with a great many wooden cubes all of the same size. In accordance with Euclidean geometry we can place them above, beside, and behind one another so as to fill a part of space of any dimensions; but this construction would never be finished; we could go on adding more and more cubes without ever finding that there was no more room. That is what we wish to express when we say that space is infinite. It would be better to say that space is infinite in relation to practically-rigid bodies, assuming that the laws of disposition for these bodies are given by Euclidean geometry.

Another example of an infinite continuum is the plane. On a plane surface we may lay squares of cardboard so that each side of any square has the side of another square adjacent to it. The construction is never finished; we can always go on laying squares – if their laws of disposition correspond to those of plane figures of Euclidean geometry. The plane is therefore infinite in relation to the cardboard squares. Accordingly we say that the plane is an infinite continuum of two dimensions, and space an infinite continuum of three dimensions. What is here meant by the number of dimensions, I think I may assume to be known.

Now we take an example of a two-dimensional continuum which is finite, but unbounded. We imagine the surface of a large globe and a quantity of small paper discs, all of the same size. We place one of the discs anywhere on the surface of the globe. If we move the disc about, anywhere we like, on the surface of the globe, we do not come upon a limit or boundary anywhere on the journey. Therefore we say that the spherical surface of the globe is an unbounded continuum. Moreover, the spherical surface is a finite continuum. For if we stick the paper discs on the globe, so that no disc overlaps another, the surface of the globe will finally become so full that there is no room for another disc. This simply means that the spherical surface of the globe is finite in relation to the paper discs. Further, the spherical surface is a non-Euclidean continuum of two dimensions, that is to say, the laws of disposition for the rigid figures lying in it do not agree with those of the Euclidean plane. This can be shown in the following way. Place a paper disc on the spherical surface, and around it in a circle place six more discs, each of which is to be surrounded in turn by six discs, and so on. If this construction is made on a plane surface, we have an uninterrupted disposition in which there are six discs touching every disc except those which lie on the outside.

On the spherical surface the construction also seems to promise success at the outset, and the smaller the radius of the disc in proportion to that of the sphere, the more promising it seems. But as the construction

Fig. 1

progresses it becomes more and more patent that the disposition of the discs in the manner indicated, without interruption, is not possible, as it should be possible by Euclidean geometry of the the plane surface. In this way creatures which cannot leave the spherical surface, and cannot even peep out from the spherical surface into three-dimensional space, might discover, merely by experimenting with discs, that their two-dimensional "space" is not Euclidean, but spherical space.

From the latest results of the theory of relativity it is probable that our three-dimensional space is also approximately spherical, that is, that the laws of disposition of rigid bodies in it are not given by Euclidean geometry, but approximately by spherical geometry, if only we consider parts of space which are sufficiently great. Now this is the place where the reader's imagination boggles. "Nobody can imagine this thing," he cries indignantly. "It can be said, but cannot be thought. I can represent to myself a spherical surface well enough, but nothing analogous to it in three dimensions."

We must try to surmount this barrier in the mind, and the patient reader will see that it is by no means a particularly difficult task. For this purpose we will first give our attention once more to the geometry of two-dimensional spherical surfaces. In the adjoining figure let K be the spherical surface, touched at S

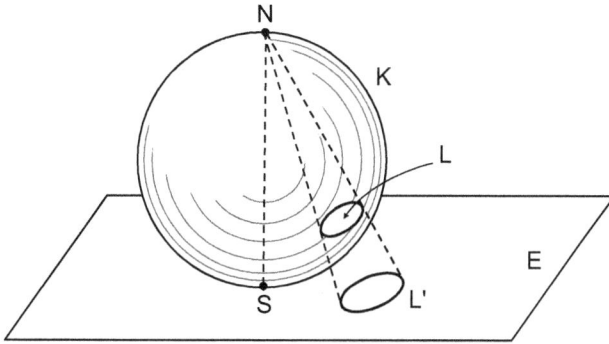

Fig. 2

by a plane, E, which, for facility of presentation, is shown in the drawing as a bounded surface. Let L be a disc on the spherical surface. Now let us imagine that at the point N of the spherical surface, diametrically opposite to S, there is a luminous point, throwing a shadow L' of the disc L upon the plane E. Every point on the sphere has its shadow on the plane. If the disc on the sphere K is moved, its shadow L' on the plane E also moves. When the disc L is at S, it almost exactly coincides with its shadow. If it moves on the spherical surface away from S upwards, the disc shadow L' on the plane also moves away from S on the plane outwards, growing bigger and bigger. As the disc L approaches the luminous point N, the shadow moves off to infinity, and becomes infinitely great.

Now we put the question, What are the laws of disposition of the disc-shadows L' on the plane E? Evidently they are exactly the same as the laws of disposition of the discs L on the spherical surface. For to each original figure on K there is a corresponding shadow figure on E. If two discs on K are touching, their shadows on E also touch. The shadow-geometry on the plane agrees with the the disc-geometry on the sphere. If we call the disc-shadows rigid figures, then spherical geometry holds good on the plane E with respect to these rigid figures. Moreover, the plane is finite with respect to the disc-shadows, since only a finite number of the shadows can find room on the plane.

At this point somebody will say, "That is nonsense. The disc-shadows are *not* rigid figures. We have only to move a two-foot rule about on the plane E to convince ourselves that the shadows constantly increase in size as they move away from S on the plane towards infinity." But what if the two-foot rule were to behave on the plane E in the same way as the disc-shadows L'? It would then be impossible to show that the shadows increase in size as they move away from S; such an assertion would then no longer have any meaning whatever. In fact the only objective assertion that can be made about the disc-shadows is just this, that they are related in exactly the same way as are the rigid discs on the spherical surface in the sense of Euclidean geometry.

We must carefully bear in mind that our statement as to the growth of the disc-shadows, as they move away from S towards infinity, has in itself no objective meaning, as long as we are unable to employ Euclidean rigid bodies which can be moved about on the plane E for the purpose of comparing the size of the disc-shadows. In respect of the laws of disposition of the shadows L', the point S has no special privileges on the plane any more than on the spherical surface.

The representation given above of spherical geometry on the plane is important for us, because it readily allows itself to be transferred to the three-dimensional case.

Let us imagine a point S of our space, and a great number of small spheres, L', which can all be brought to coincide with one another. But these spheres are not to be rigid in the sense of Euclidean geometry; their radius is to increase (in the sense of Euclidean geometry) when they are moved away from S towards infinity, and this increase is to take place in exact accordance with the same law as applies to the increase of the radii of the disc-shadows L' on the plane.

After having gained a vivid mental image of the geometrical behaviour of our L' spheres, let us assume that in our space there are no rigid bodies at all in the sense of Euclidean geometry, but only bodies having the behaviour of our L' spheres. Then we shall have a vivid representation of three-dimensional spherical space, or, rather of three-dimensional spherical geometry. Here our spheres must be called "rigid" spheres. Their increase in size as they depart from S is not to be detected by measuring with measuring-rods, any more than in the case of the disc-shadows on E, because the standards of measurement behave in the same way as the spheres. Space is homogeneous, that is to say, the same spherical configurations are possible in the environment of all points.[1] Our space is finite, because, in consequence of the "growth" of the spheres, only a finite number of them can find room in space.

In this way, by using as stepping-stones the practice in thinking and visualisation which Euclidean geometry gives us, we have acquired a mental picture of spherical geometry. We may without difficulty impart more depth and vigour to these ideas by carrying

---

[1] This is intelligible without calculation – but only for the two-dimensional case – if we revert once more to the case of the disc on the surface of the sphere.

67

out special imaginary constructions. Nor would it be difficult to represent the case of what is called elliptical geometry in an analogous manner. My only aim today has been to show that the human faculty of visualisation is by no means bound to capitulate to non-Euclidean geometry.

THE EINSTEIN THEORY OF RELATIVITY[*]

A CONCISE STATEMENT

BY

PROF. H. A. LORENTZ

of the University of Leyden

[*]H. A. Lorentz, *The Einstein Theory of Relativity: A Concise Statement*, 3rd ed. (Brentano's Publishers, New York 1920). Originally published in the *Nieuwe Rotterdamsche Courant* of November 19, 1919.

69

NOTE

Whether it is true or not that not more than twelve persons in all the world are able to understand Einstein's Theory, it is nevertheless a fact that there is a constant demand for information about this much-debated topic of relativity. The books published on the subject are so technical that only a person trained in pure physics and higher mathematics is able to fully understand them. In order to make a popular explanation of this far-reaching theory available, the present book is published.

Professor Lorentz is credited by Einstein with sharing the development of his theory. He is doubtless better able than any other man – except the author himself – to explain this scientific discovery.

The publishers wish to acknowledge their indebtedness to the New York *Times*, *The Review of Reviews* and *The Athenaeum* for courteous permission to reprint articles from their pages. Professor Lorentz's article appeared originally in *The Nieuwe Rotterdamsche Courant* of November 19, 1919.

INTRODUCTION

The action of the Royal Society at its meeting in London on November 6, in recognizing Dr. Albert Einstein's "theory of relativity" has caused a great stir in scientific circles on both sides of the Atlantic. Dr. Einstein propounded his theory nearly fifteen years ago. The present revival of interest in it is due to the remarkable confirmation which it received in the report of the observations made during the sun's eclipse of last May to determine whether rays of light passing close to the sun are deflected from their course.

The actual deflection of the rays that was discovered by the astronomers was precisely what had been predicted theoretically by Einstein many years since. This striking confirmation has led certain German scientists to assert that no scientific discovery of such importance has been made since Newton's theory of gravitation was promulgated. This suggestion, however, was put aside by Dr. Einstein himself when he was interviewed by a correspondent of the New York *Times* at his home in Berlin. To this correspondent he expressed the difference between his conception and the law of gravitation in the following terms:

"Please imagine the earth removed, and in its place suspended a box as big as a room or a whole house, and inside a man naturally floating in the center, there being no force whatever pulling him. Imagine, further, this box being, by a rope or other contrivance, suddenly jerked to one side, which is scientifically termed 'difform motion', as opposed to 'uniform motion.' The person would then naturally reach bottom on the opposite side. The result would consequently be the same as if he obeyed Newton's law of gravitation, while, in fact, there is no gravitation exerted whatever, which proves that difform motion will in every case produce the same effects as gravitation.

"I have applied this new idea to every kind of difform motion and have thus developed mathematical formulas which I am convinced give more precise results than those based on Newton's theory. Newton's formulas, however, are such close approximations that it was difficult to find by observation any obvious disagreement with experience."

Dr. Einstein, it must be remembered, is a physicist and not an astronomer. He developed his theory as a mathematical formula. The confirmation of it came from the astronomers. As he himself says, the crucial test was supplied by the last total solar eclipse. Observations then proved that the rays of fixed stars, having to pass close to the sun to reach the earth, were deflected the exact amount demanded by Einstein's formulas. The deflection was also in the direction predicted by him.

The question must have occurred to many, what has all this to do with relativity? When this query was propounded by the *Times* correspondent to Dr. Einstein he replied as follows:

"The term relativity refers to time and space. According to Galileo and Newton, time and space were absolute entities, and the moving systems of the universe were dependent on this absolute time and space.

On this conception was built the science of mechanics. The resulting formulas sufficed for all motions of a slow nature; it was found, however, that they would not conform to the rapid motions apparent in electrodynamics.

This led the Dutch professor, Lorentz, and myself to develop the theory of special relativity. Briefly, it discards absolute time and space and makes them in every instance relative to moving systems. By this theory all phenomena in electrodynamics, as well as mechanics, hitherto irreducible by the old formulae – and there are multitudes – were satisfactorily explained.

Till now it was believed that time and space existed by themselves, even if there was nothing else – no sun, no earth, no stars – while now we know that time and space are not the vessel for the universe, but could not exist at all if there were no contents,[1] namely, no sun, earth and other celestial bodies.

This special relativity, forming the first part of my theory, relates to all systems moving with uniform motion; that is, moving in a straight line with equal velocity.[2]

---

[1]EDITOR'S NOTE: This position appears to be a result of the influence of Mach's ideas on Einstein; This is not the case as can be seen in Minkowski's 1908 lecture "Space and Time" and also from de Sitter's solution of Einstein's equation which presents such a universe that is empty of matter.

[2]EDITOR'S NOTE: Probably this early explanations by Einstein had given rise to the misconception that acceleration is not treated in special relativity. Apparently Einstein did not pay attention that his mathematics professor – Hermann Minkowski – demonstrated in his 1908 groundbreaking lecture "Space and Time" that acceleration is a natural feature of flat spacetime (where Einstein's special relativity holds) [1]:

Having individualized space and time in some way, a straight worldline parallel to the t-axis

Gradually I was led to the idea, seeming a very paradox in science, that it might apply equally to all moving systems, even of difform motion, and thus I developed the conception of general relativity which forms the second part of my theory."

As summarized by an American astronomer, Professor Henry Norris Russell, of Princeton, in the *Scientific American* for November 29, Einstein's contribution amounts to this:

"The central fact which has been proved – and which is of great interest and importance – is that the natural phenomena involving gravitation and inertia (such as the motions of the planets) and the phenomena involving electricity and magnetism (including the motion of light) are not independent of one another, but are intimately related, so that both sets of phenomena should be regarded as parts of one vast system, embracing all Nature. The relation of the two is, however, of such a character that it is perceptible only in a very few instances, and then only to refined observations."

Already before the war, Einstein had immense fame among physicists, and among all who are interested in the philosophy of science, because of his principle of relativity.

---

corresponds to a stationary substantial point, a straight line inclined to the t-axis corresponds to a uniformly moving substantial point, a somewhat curved worldline corresponds to a non-uniformly moving substantial point.

And also on p. 66: "Especially the concept of *acceleration* acquires a sharply prominent character."

1. H. Minkowski, "Space and Time" in: H. Minkowski, *Spacetime: Minkowski's Papers on Spacetime Physics* (Minkowski Institute Press, Montreal 2020), pp. 57-76, p. 62.

Clerk Maxwell had shown that light is electromagnetic, and had reduced the whole theory of electromagnetism to a small number of equations, which are fundamental in all subsequent work. But these equations were entangled with the hypothesis of the ether, and with the notion of motion relative to the ether. Since the ether was supposed to be at rest, such motion was indistinguishable from absolute motion. The motion of the earth relatively to the ether should have been different at different points of its orbit, and measurable phenomena should have resulted from this difference. But none did, and all attempts to detect effects of motions relative to the ether failed. The theory of relativity succeeded in accounting for this fact. But it was necessary incidentally to throw over the one universal time, and substitute local times attached to moving bodies and varying according to their motion. The equations on which the theory of relativity is based are due to Lorentz, but Einstein connected them with his general principle, namely, that there must be nothing, in observable phenomena, which could be attributed to absolute motion of the observer.

In orthodox Newtonian dynamics the principle of relativity had a simpler form, which did not require the substitution of local time for general time. But it now appeared that Newtonian dynamics is only valid when we confine ourselves to velocities much less than that of light. The whole Galileo-Newton system thus sank to the level of a first approximation, becoming progressively less exact as the velocities concerned approached that of light.

Einstein's extension of his principle so as to account for gravitation was made during the war, and for a considerable period our astronomers were unable to become acquainted with it, owing to the difficulty of obtaining German printed matter. However, copies of his work ultimately reached the outside world and enabled people to learn more about it. Gravitation, ever since Newton, had remained isolated from other forces in nature; various attempts had been made to account for it, but without success. The immense unification effected by electro-magnetism apparently left gravitation out of its scope. It seemed that nature had presented a challenge to the physicists which none of them were able to meet.

At this point Einstein intervened with a hypothesis which, apart altogether from subsequent verification, deserves to rank as one of the great monuments of human genius. After correcting Newton, it remained to correct Euclid, and it was in terms of non-Euclidean geometry that he stated his new theory. Non-Euclidean geometry is a study of which the primary motive was logical and philosophical; few of its promoters ever dreamed that it would come to be applied in physics. Some of Euclid's axioms were felt to be not "necessary truths," but mere empirical laws; in order to establish this view, self-consistent geometries were constructed upon assumptions other than those of Euclid. In these geometries the sum of the angles of a triangle is not two right angles, and the departure from two right angles increases as the size of the triangle increases. It is often said that in non-Euclidean geometry space has a curvature, but this way of stating the matter is misleading, since it seems to imply a fourth dimension, which is not implied by these systems.

Einstein supposes that space is Euclidean where it is sufficiently remote from matter, but that the presence of matter causes it to become slightly non-Euclidean – the more matter there is in the neighborhood, the more space will depart from Euclid. By the help of this hypothesis, together with his previous theory of relativity, he deduces gravitation – very approximately, but not exactly, according to the Newtonian law of the inverse square. The minute differences between the effects deduced from his theory and those deduced from Newton are measurable in certain cases. There are, so far, three crucial tests of the relative accuracy of the new theory and the old.

(1) The perihelion of Mercury shows a discrepancy which has long puzzled astronomers. This discrepancy is fully accounted for by Einstein. At the time when he published his theory, this was its only experimental verification.

(2) Modern physicists were willing to suppose that light might be subject to gravitation – i.e., that a ray of light passing near a great mass like the sun might be deflected to the extent to which a particle moving with the same velocity would be deflected according to the orthodox theory of gravitation. But Einstein's theory required that the light should be deflected just twice as much as this. The matter could only be tested during an eclipse among a number of bright stars. Fortunately a peculiarly favourable eclipse occurred last year. The results of the observations have now been published, and are found to verify Einstein's prediction. The verification is not, of course, quite exact; with such delicate observations that was not to be expected. In some cases the departure is considerable. But taking the average of the best series of observations, the deflection at the sun's limb is found to be 1.98″, with a probable error of about 6 per cent., whereas the deflection calculated by Einstein's theory should be 1.75″. It will be noticed that Einstein's theory gave a deflection twice as large as that predicted

by the orthodox theory, and that the observed deflection is slightly *larger* than Einstein predicted. The discrepancy is well within what might be expected in view of the minuteness of the measurements. It is therefore generally acknowledged by astronomers that the outcome is a triumph for Einstein.

(3) In the excitement of this sensational verification, there has been a tendency to overlook the third experimental test to which Einstein's theory was to be subjected. If his theory is correct as it stands, there ought, in a gravitational field, to be a displacement of the lines of the spectrum towards the red. No such effect has been discovered.[3] Spectroscopists maintain that, so far as can be seen at present, there is no way of accounting for this failure if Einstein's theory in its present form is assumed. They admit that some compensating cause *may* be discovered to explain the discrepancy, but they think it far more probable that Einstein's theory requires some essential modification. Meanwhile, a certain suspense of judgment is called for. The new law has been so amazingly successful in two of the three tests that there must be some thing valid about it, even if it is not exactly right as yet.

Einstein's theory has the very highest degree of aesthetic merit: every lover of the beautiful must wish it to be true. It gives a vast unified survey of the operations of nature, with a technical simplicity in the critical assumptions which makes the wealth of deductions astonishing. It is a case of an advance arrived at by pure theory: the whole effect of Einstein's work is to make physics more philosophical (in a good sense), and to restore some of that intellectual unity which belonged to the great scientific systems of the seven-

---

[3]EDITOR'S NOTE: See Editor's note on p. 6.

teenth and eighteenth centuries, but which was lost through increasing specialization and the overwhelming mass of detailed knowledge. In some ways our age is not a good one to live in, but for those who are interested in physics there are great compensations.

The Einstein Theory of Relativity

A Concise Statement by Prof. H. A. Lorentz of the University of Leyden

The total eclipse of the sun of May 29, resulted in a striking confirmation of the new theory of the universal attractive power of gravitation developed by Albert Einstein, and thus reinforced the conviction that the defining of this theory is one of the most important steps ever taken in the domain of natural science. In response to a request by the editor, I will attempt to contribute something to its general appreciation in the following lines.

For centuries Newton's doctrine of the attraction of gravitation has been the most prominent example of a theory of natural science. Through the simplicity of its basic idea, an attraction between two bodies proportionate to their mass and also proportionate to the square of the distance; through the completeness with which it explained so many of the peculiarities in the movement of the bodies making up the solar system; and, finally, through its universal validity, even in the case of the far-distant planetary systems, it compelled the admiration of all.

But, while the skill of the mathematicians was devoted to making more exact calculations of the consequences to which it led, no real progress was made in the science of gravitation. It is true that the inquiry was transferred to the field of physics, following Cavendish's success in demonstrating the common attraction between bodies with which laboratory work can be done, but it always was evident that natural philosophy had no grip on the universal power of attraction. While in electric effects an influence exercised by the matter placed between bodies was speedily observed – the starting-point of a new and fertile doctrine of electricity – in the case of gravitation not a trace of an influence exercised by intermediate matter could ever be discovered. It was, and remained, inaccessible and unchangeable, without any connection, apparently, with other phenomena of natural philosophy.

Einstein has put an end to this isolation; it is now well established that gravitation affects not only matter, but also light. Thus strengthened in the faith that his theory already has inspired, we may assume with him that there is not a single physical or chemical phenomenon – which does not feel, although very probably in an unnoticeable degree, the influence of gravitation, and that, on the other side, the attraction exercised by a body is limited in the first place by the quantity of matter it contains and also, to some degree, by motion and by the physical and chemical condition in which it moves.

It is comprehensible that a person could not have arrived at such a far-reaching change of view by continuing to follow the old beaten paths, but only by introducing some sort of new idea. Indeed, Einstein arrived at his theory through a train of thought of great originality. Let me try to restate it in concise terms.

THE EARTH AS A MOVING CAR

Everyone knows that a person may be sitting in any kind of a vehicle without noticing its progress, so long as the movement does not vary in direction or speed; in a car of a fast express train objects fall in just the same way as in a coach that is standing still. Only when we look at objects outside the train, or when the air can enter the car, do we notice indications of the motion. We may compare the earth with such a moving vehicle, which in its course around the sun has a remarkable speed, of which the direction and velocity during a considerable period of time may be regarded as constant. In place of the air now comes, so it was reasoned formerly, the ether which fills the spaces of the universe and is the carrier of light and of electromagnetic phenomena; there were good reasons to assume that the earth was entirely permeable for the ether and could travel through it without setting it in motion. So here was a case comparable with that of a railroad coach open on all sides. There certainly should have been a powerful "ether wind" blowing through the earth and all our instruments, and it was to have been expected that some signs of it would be noticed in connection with some experiment or other. Every attempt along that line, however, has remained fruitless;

all the phenomena examined were evidently independent of the motion of the earth. That this is the way they do function was brought to the front by Einstein in his first or "special" theory of relativity. For him the ether does not function and in the sketch that he draws of natural phenomena there is no mention of that intermediate matter.

If the spaces of the universe are filled with an ether, let us suppose with a substance, in which, aside from eventual vibrations and other slight movements, there is never any crowding or flowing of one part alongside of another, then we can imagine fixed points existing in it; for example, points in a straight line, located one meter apart, points in a level plain, like the angles or squares on a chess board extending out into infinity, and finally, points in space as they are obtained by repeatedly shifting that level spot a distance of a meter in the direction perpendicular to it. If, consequently, one of the points is chosen as an "original point" we can, proceeding from that point, reach any other point through three steps in the common perpendicular directions in which the points are arranged. The figures showing how many meters are comprized in each of the steps may serve to indicate the place reached and to distinguish it from any other; these are, as is said, the "coordinates" of these places, comparable, for example, with the numbers on a map giving the longitude and latitude. Let us imagine that each point has noted upon it the three numbers that give its position, then we have something comparable with a measure with numbered subdivisions; only we now have to do, one might say, with a good many imaginary measures in three common perpendicular directions. In this "system of coordinates" the numbers that fix the position of one or the other of the bodies may now be read off at any moment.

This is the means which the astronomers and their mathematical assistants have always used in dealing with the movement of the heavenly bodies. At a determined moment the position of each body is fixed by its three coordinates. If these are given, then one knows also the common distances, as well as the angles formed by the connecting lines, and the movement of a planet is to be known as soon as one knows how its coordinates are changing from one moment to the other. Thus the picture that one forms of the phenomena stands there as if it were sketched on the canvas of the motionless ether.

EINSTEIN'S DEPARTURE

Since Einstein has cut loose from the ether, he lacks this canvas, and therewith, at the first glance, also loses the possibility of fixing the positions of the heavenly bodies and mathematically describing their movement – i.e., by giving comparisons that define the positions at every moment. How Einstein has overcome this difficulty may be somewhat elucidated through a simple illustration.

On the surface of the earth the attraction of gravitation causes all bodies to fall along vertical lines, and, indeed, when one omits the resistance of the air, with an equally accelerated movement; the velocity increases in equal degrees in equal consecutive divisions of time at a rate that in this country gives the velocity attained at the end of a second as 981 centimeters (32.2 feet) per second. The number 981 defines the "acceleration in the field of gravitation," and this field is fully characterized by that single number; with its help we can also calculate the movement of an object hurled out in an arbitrary direction. In order to measure the acceleration we let the body drop alongside of a vertical measure set solidly on the ground; on this scale we read at every moment the figure that indicates the height, the only coordinate that is of importance in this rectilinear movement. Now we ask what would we be able to see if the measure were not bound solidly to the earth, if it, let us suppose, moved down or up with the place where it is located and where we are ourselves. If in this case the speed were constant, then, and this is in accord with the special theory of relativity, there would be no motion observed at all; we should again find an acceleration of 981 for a falling body. It would be different if the measure moved with changeable velocity.

If it went down with a constant acceleration of 981 itself, then an object could remain permanently at the same point on the measure, or could move up or down itself alongside of it, with constant speed. The relative movement of the body with regard to the measure should be without acceleration, and if we had to judge only by what we observed in the spot where we were and which was falling itself, then we should get the impression that there was no gravitation at all. If the measure goes down with an acceleration equal to a half or a third of what it just was, then the relative motion of the body will, of course, be accelerated, but we should find the increase in velocity per second one-half or two-thirds of 981. If, finally, we let the measure rise with a uniformly accelerated movement, then we shall find a greater acceleration than 981 for the body itself.

Thus we see that we, also when the measure is not attached to the earth, disregarding its displacement, may describe the motion of the body in respect to the measure always in the same way – i.e., as one uniformly accelerated, as we ascribe now and again a fixed value to the acceleration of the sphere of gravitation, in a particular case the value of zero.

Of course, in the case here under consideration the use of a measure fixed immovably upon the earth should merit all recommendation. But in the spaces of the solar system we have, now that we have abandoned the ether, no such support. We can no longer establish a system of coordinates, like the one just mentioned, in a universal intermediate matter, and if we were to arrive in one way or another at a definite system of lines crossing each other in three directions, then we should be able to use just as well another similar system that in respect to the first moves this or that way. We should also be able to remodel the system of coordinates in all kinds of ways, for example by extension or compression. That in all these cases for fixed bodies that do not participate in the movement or the remodelling of the system other coordinates will be read off again and again is clear.

New System of
Coordinates

What way Einstein had to follow is now apparent. He must – this hardly needs to be said – in calculating definite, particular cases make use of a chosen system of coordinates, but as he had no means of limiting his choice beforehand and in general, he had to reserve full liberty of action in this respect. Therefore he made it his aim so to arrange the theory that, no matter how the choice was made, the phenomena of gravitation, so far as its effects and its stimulation by the attracting bodies are concerned, may always be described in the same way – i.e., through comparisons of the same general form, as we again and again give certain values to the numbers that mark the sphere of gravitation. (For the sake of simplification I here disregard the fact that Einstein desires that also the way in which time is measured and represented by figures shall have no influence upon the central value of the comparisons.)

Whether this aim could be attained was a question of mathematical inquiry. It really was attained, remarkably enough, and, we may say, to the surprise of Einstein himself, although at the cost of considerable simplicity in the mathematical form; it appeared necessary for the fixation of the field of gravitation in one or the other point in space to introduce no fewer than ten quantities in the place of the one that occurred in the example mentioned above.

In this connection it is of importance to note that when we exclude certain possibilities that would give rise to still greater intricacy, the form of comparison used by Einstein to present the theory is the only possible one; the principle of the freedom of choice in coordinates was the only one by which he needed to allow himself to be guided. Although thus there was no special effort made to reach a connection with the theory of Newton, it was evident, fortunately, at the end of the experiment that the connection existed. If we avail ourselves of the simplifying circumstance that the velocities of the heavenly bodies are slight in comparison with that of light, then we can deduce the theory of Newton from the new theory, the "universal" relativity theory, as it is called by Einstein. Thus all the conclusions based upon the Newtonian theory hold good, as must naturally be required. But now we have got further along. The Newtonian theory can no longer be regarded as absolutely correct in all cases; there are slight deviations from it, which, although as a rule unnoticeable, once in a while fall within the range of observation.

Now, there was a difficulty in the movement of the planet Mercury which could not be solved. Even after all the disturbances caused by the attraction of other planets had been taken into account, there remained an inexplicable phenomenon – i.e., an extremely slow turning of the ellipsis described by Mercury on its own plane; Leverrier had found that it amounted to forty-three seconds a century. Einstein found that, according to his formulas, this movement must really amount to just that much. Thus with a single blow he solved one of the greatest puzzles of astronomy.

Still more remarkable, because it has a bearing upon a phenomenon which formerly could not be imagined, is the confirmation of Einstein's prediction regarding the influence of gravitation upon the course of the rays of light. That such an influence must exist is taught by a simple examination; we have only to turn back for a moment to the following comparison in which we were just imagining ourselves to make our observations. It was noted that when the compartment is falling with the acceleration of 981 the phenomena therein will occur just as if there were no attraction of gravitation. We can then see an object, A, stand still somewhere in open space. A projectile, B, can travel with constant speed along a horizontal line, without varying from it in the slightest.

A ray of light can do the same; everybody will admit that in each case, if there is no gravitation, light will certainly extend itself in a rectilinear way. If we limit the light to a flicker of the slightest duration, so that only a little bit, C, of a ray of light arises, or if we fix our attention upon a single vibration of light, C, while we on the other hand give to the projectile, B, a speed equal to that of light, then we can conclude that B and C in their continued motion can always remain next to each other. Now if we watch all this, not from the movable compartment, but from a place on the earth, then we shall note the usual falling movement of object A, which shows us that we have to deal with a sphere of gravitation. The projectile B will, in a bent path, vary more and more from a horizontal straight line, and the light will do the same, because if we observe the movements from another standpoint this can have no effect upon the remaining next to each other of B and C.

DEFLECTION OF LIGHT

The bending of a ray of light thus described is much too light on the surface of the earth to be observed. But the attraction of gravitation exercised by the sun on its surface is, because of its great mass, more than twenty-seven times stronger, and a ray of light that goes close by the superficies of the sun must surely be noticeably bent. The rays of a star that are seen at a short distance from the edge of the sun will, going along the sun, deviate so much from the original direction that they strike the eye of an observer as if they came in a straight line from a point somewhat further removed than the real position of the star from the sun. It is at that point that we think we see the star; so here is a seeming displacement from the sun, which increases in the measure in which the star is observed closer to the sun. The Einstein theory teaches that the displacement is in inverse proportion to the apparent distance of the star from the centre of the sun, and that for a star just on its edge it will amount to $1'.75$ (1.75 seconds). This is approximately the thousandth part of the apparent diameter of the sun.

Naturally, the phenomenon can only be observed when there is a total eclipse of the sun; then one can take photographs of neighboring stars and through comparing the plate with a picture of the same part of the heavens taken at a time when the sun was far removed from that point the sought-for movement to one side may become apparent.

Thus to put the Einstein theory to the test was the principal aim of the English expeditions sent out to observe the eclipse of May 29, one to Prince's Island, off the coast of Guinea, and the other to Sobral, Brazil. The first-named expedition's observers were Eddington and Cottingham, those of the second, Crommelin and Davidson. The conditions were especially favorable, for a very large number of bright stars were shown on the photographic plate; the observers at Sobral being particularly lucky in having good weather.

The total eclipse lasted five minutes, during four of which it was perfectly clear, so that good photographs could be taken. In the report issued regarding the results the following figures, which are the average of the measurements made from the seven plates, are given for the displacements of seven stars:

$1''.02$, $0''.92$, $0''.84$, $0''.58$, $0''.54$, $0''.36$, $0''.24$, whereas, according to the theory, the displacements should have amounted to: $0''.88$, $0''.80$, $0''.75$, $0''.40$, $0''.52$, $0''.33$, $0''.20$.

If we consider that, according to the theory the displacements must be in inverse ratio to the distance from the centre of the sun, then we may deduce from each observed displacement how great the sideways movement for a star at the edge of the sun should have been. As the most probable result, therefore, the number $1''.98$ was found from all the observations together. As the last of the displacements given above – i.e., $0''.24$ is about one-eighth of this, we may say that the influence of the attraction of the sun upon light made itself felt upon the ray at a distance eight times removed from its centre.

The displacements calculated according to the theory are, just because of the way in which they are calculated, in inverse proportion to the distance to the centre. Now that the observed deviations also accord with the same rule, it follows that they are surely proportionate with the calculated displacements. The proportion of the first and the last observed sidewise movements is 4.2, and that of the two most extreme of the calculated numbers is 4.4.

This result is of importance, because thereby the theory is excluded, or at least made extremely improbable, that the phenomenon of refraction is to be ascribed to, a ring of vapor surrounding the sun for a great distance. Indeed, such a refraction should cause a deviation in the observed direction, and, in order to produce the displacement of one of the stars under observation itself a slight proximity of the vapor ring should be sufficient, but we have every reason to expect that if it were merely a question of a mass of gas around the sun the diminishing effect accompanying a removal from the sun should manifest itself much faster than is really the case. We cannot speak with perfect certainty here, as all the factors that might be of influence upon the distribution of density in a sun atmosphere are not well enough known, but we can surely demonstrate that in case one of the gasses with which we are acquainted were held in equilibrium solely by the influence of attraction of the sun the phenomenon should become much less as soon as we got somewhat further from the edge of the sun. If the displacement of the first star, which amounts to 1.02-seconds were to be ascribed to such a mass of gas, then the displacement of the second must already be entirely inappreciable.

So far as the absolute extent of the displacements is concerned, it was found somewhat too great, as has been shown by the figures given above; it also appears from the final result to be 1.98 for the edge of the sun – i.e., 13 per cent, greater than the theoretical value of 1.75. It indeed seems that the discrepancies may be ascribed to faults in observations, which supposition is supported by the fact that the observations at Prince's Island, which, it is true, did not turn out quite as well as those mentioned above, gave the result, of 1.64, somewhat lower than Einstein's figure.

(The observations made with a second instrument at Sobral gave a result of 0.93, but the observers are of the opinion that because of the shifting of the mirror which reflected the rays no value is to be attached to it.)

DIFFICULTY EXAGGERATED

During a discussion of the results obtained at a joint meeting of the Royal Society and the Royal Astronomical Society held especially for that purpose recently in London, it was the general opinion that Einstein's prediction might be regarded as justified, and warm tributes to his genius were made on all sides. Nevertheless, I cannot refrain, while I am mentioning it, from expressing my surprise that, according to the report in *The Times* there should be so much complaint about the difficulty of understanding the new theory. It is evident that Einstein's little book "About the Special and the General Theory of Relativity in Plain Terms," did not find its way into England during wartime. Any one reading it will, in my opinion, come to the conclusion that the basic ideas of the theory are really clear and simple; it is only to be regretted that it was impossible to avoid clothing them in pretty involved mathematical terms, but we must not worry about that.

I allow myself to add that, as we follow Einstein, we may retain much of what has been formerly gained. The Newtonian theory remains in its full value as the first great step, without which one cannot imagine the development of astronomy and without which the second step, that has now been made, would hardly have been possible. It remains, moreover, as the first, and in most cases, sufficient, approximation. It is true that, according to Einstein's theory, because it leaves us entirely free as to the way in which we wish to represent the phenomena, we can imagine an idea of the solar system in which the planets follow paths of peculiar form and the rays of light shine along sharply bent lines – think of a twisted and distorted planetarium – but in every case where we apply it to concrete questions we shall so arrange it that the planets describe almost exact ellipses and the rays of light almost straight lines.

It is not necessary to give up entirely even the ether.[4] Many natural philosophers find satisfaction in the idea of a material intermediate substance in which the vibrations of light take place, and they will very probably be all the more inclined to imagine such a medium when they learn that, according to the Einstein theory, gravitation itself does not spread instantaneously, but with a velocity that at the first estimate may be compared with that of light. Especially in former years were such interpretations current and repeated attempts were made by speculations about the nature of the ether and about the mutations and movements that might take place in it to arrive at a clear presentation of electromagnetic phenomena, and also of the functioning of gravitation. In my opinion it is not impossible that in the future this road, indeed abandoned at present, will once more be followed with good results, if only because it can lead to the thinking out of new experimental tests. Einstein's theory need not keep us from so doing; only the ideas about the ether must accord with it.

---

[4]Editor's Note: See Editor's note on p. 50.

Nevertheless, even without the color and clearness that the ether theories and the other models may be able to give, and even, we can feel it this way, just because of the soberness induced by their absence, Einstein's work, we may now positively expect, will remain a monument of science; his theory entirely fulfills the first and principal demand that we may make, that of deducing the course of phenomena from certain principles exactly and to the smallest details. It was certainly fortunate that he himself put the ether in the background; if he had not done so, he probably would never have come upon the idea that has been the foundation of all his examinations.

Thanks to his indefatigable exertions and perseverance, for he had great difficulties to overcome in his attempts, Einstein has attained the results, which I have tried to sketch, while still young; he is now 45 years old. He completed his first investigations in Switzerland, where he first was engaged in the Patent Bureau at Berne and later as a professor at the Polytechnic in Zurich. After having been a professor for a short time at the University of Prague, he settled in Berlin, where the Kaiser Wilhelm Institute afforded him the opportunity to devote himself exclusively to his scientific work. He repeatedly visited our country and made his Netherland colleagues, among whom he counts many good friends, partners in his studies and his results. He attended the last meeting of the department of natural philosophy of the Royal Academy of Sciences, and the members then had the privilege of hearing him explain, in his own fascinating, clear and simple way, his interpretations of the fundamental questions to which his theory gives rise.

INDEX

www.ingramcontent.com/pod-product-compliance
Lightning Source LLC
Chambersburg PA
CBHW071454200326
41519CB00019B/5727